THE ULTIMATE GUIDE TO SPAS AND HOT TUBS

Terry Tamminen

McGraw-Hill

New York   Chicago   San Francisco   Lisbon   London   Madrid
Mexico City   Milan   New Delhi   San Juan   Seoul
Singapore   Sydney   Toronto

The McGraw·Hill Companies

Cataloging-in-Publication Data is on file with the Library of Congress

1 2 3 4 5 6 7 8 9 0   DOC/DOC   0 1 0 9 8 7 6 5

ISBN 0-07-143921-8

The sponsoring editor for this book was Larry S. Hager, the editing supervisor was Stephen M. Smith, and the production supervisors were Sherri Souffrance and Richard Ruzycka. It was set in Melior by Wayne A. Palmer of McGraw-Hill Professional's Hightstown, N.J., composition unit. The art director for the cover was Anthony Landi.

Printed and bound by RR Donnelley.

McGraw-Hill books are available at special quantity discounts to use as premiums and sales promotions, or for use in corporate training programs. For more information, please write to the Director of Special Sales, McGraw-Hill Professional, Two Penn Plaza, New York, NY 10121-2298. Or contact your local bookstore.

 This book is printed on recycled, acid-free paper containing a minimum of 50% recycled, de-inked fiber.

To Bonnie Reiss, Robert F. Kennedy, Jr., Maria Shriver, and Arnold Schwarzenegger—for all of the hot water we have been in together...and all that is yet to come.

CONTENTS

ACKNOWLEDGMENTS

There are literally hundreds of people who have helped make this book possible, including my many residential and commercial pool and spa clients over the years. (Yes, we professionals learn a lot from our clients too!) Since these people are too numerous to name, they should please accept this book, and their contributions in it, as my thanks.

The products featured in this book are meant to illustrate various components of spas and hot tubs, but there are many fine comparable products that could just as easily been shown. Search your pool/spa retailer and the Web for more variety in style, function, and price in every category. You will be pleasantly surprised about your choices.

Some people and companies deserve a special thanks, particularly those who allowed the use of their technical information and illustrations throughout the book. Thank you to my partner of two decades, Ritchie Creevy, and thank you to

- Aqua-Flo, Inc., Chino, California
- Bradford Spas, LLC, Wilmington, North Carolina
- California Cooperage of Santa Monica, California, a Jacuzzi Premium dealer
- Confer Plastics, North Tonawanda, New York
- Diamond Spas, Broomfield, Colorado
- Essentials, a division of Spires Management, Inc., Cumming, Georgia
- Fafco, Chico, California
- Fiber Optic Technologies, Mississauga, Ontario, Canada
- Gordon & Grant Hot Tubs, Santa Barbara, California

- Great Northern Hot Tubs, Minneapolis, Minnesota
- Jacuzzi, Little Rock, Arkansas
- Jandy, Petaluma, California
- Master Spas, Ft. Wayne, Indiana
- Medallion Pools, Matoaca, Virginia
- Nordic Hot Tubs, Grand Rapids, Michigan
- Pentair Pool Products, Inc., Moorpark, California
- Poolmaster, Inc., Sacramento, California
- Raypak, Inc., Oxnard, California
- Sta-Rite Industries, Delevan, Wisconsin
- Sundance Spas, Chino, California
- Sunstar Enterprises, San Marcos, California
- Suntrek Industries, Laguna Hills, California

INTRODUCTION

Some years ago, I began working with local environmental groups to restore Santa Monica Bay and its famed beaches. One of my pool clients, actor Dustin Hoffman, heard about the effort and volunteered to help, saying "It's about time you cleaned up the BIG pool!"

Whether your pool seems as big as the Pacific Ocean or is a modest portable spa, everyone can enjoy clean, healthy bathing water. Like my books *The Ultimate Pool Maintenance Manual* and *The Ultimate Guide to Above-Ground Pools*, this book will teach you the secrets of keeping water sparkling clean and easy to maintain. In over 30 years of hands-on experience, I have collected the tips of the old pros and the newest high-technology concepts from around the world. Now you can benefit from this expert advice, whether you are a service technician yourself or just a weekend do-it-yourselfer.

My Malibu and Beverly Hills clients have included Barbra Streisand, Dick Van Dyke, Madonna, and super-agent Michael Ovitz, but by following the procedures and tips outlined in this book, you can enjoy the same spa experience as the rich and famous. Reflecting the rapid advances in the pool and spa industry over the past decade, *The Ultimate Guide to Spas and Hot Tubs* contains information presented in no other book about pools and spas, including

- Easy, Advanced, or Pro ratings for all procedures so you know what you might tackle and when to call the experts.

- Cutting-edge technologies such as fiber optic lighting and solid-state remote controls.

- Easy-to-read troubleshooting charts.

- Measurements and formulas in both metric and U.S. standard systems.

■ Details and repair tips about the fastest-growing segment of the pool industry—portable spas.

■ Solar heating made easy.

■ Tools of the Trade checklists to make preparation for repair jobs easier.

■ The latest on health and safety issues for residential and commercial spas.

■ A checklist of innovative things your spa can do for our environment.

■ Great remodeling techniques for spas.

■ Chlorine alternatives, such as practical high-tech choices like ozone.

■ Useful, "cool" sites on the Web to keep you up to date with the latest developments in spa maintenance and accessories.

■ Genuine Tricks of the Trade, in print here for the first time anywhere. These insights can make the difference between success and frustration in spa maintenance, water chemistry, and equipment repairs.

If you are considering the purchase of a new spa or hot tub, this book will also save you time and money, and help you make the right choice for your needs. An investment in a spa or hot tub can cost as little as $500 or as much as $50,000, but no matter the price or amenities, your installation can provide decades of use and enjoyment by the whole family if properly maintained. *The Ultimate Guide to Spas and Hot Tubs* is designed to help you do just that.

Come on in—the water's fine!

Terry Tamminen

The Spa and Hot Tub

Nature designed the first therapeutic spa, and for thousands of years, humans and animals alike have enjoyed the benefits of hot, bubbling water.

Monkeys in the coldest winters in Japan seek refuge in soothing hot water (Fig. 1-1). Sulfur and mineral hot springs abound throughout the world, nowhere more dramatically than in Yellowstone National Park (Fig. 1-2).

But modern humans have perfected what nature invented, recreating the hot spring experience in homes, health clubs, and just about anywhere that people gather for recreation. We have brought nature into our lives, using the latest space-age materials and high-tech controls, but we have also embraced the simple elegance of bubbling hot water in a tub made of cedar or redwood.

Today, spas and hot tubs are made in hundreds of different shapes, sizes, and formats—truly something for every need and pocketbook. Modern plastics allow form-fitted shapes for every body with tailored jets to provide the maximum relaxation and enjoyment. Spas are even made of lightweight foam materials that can be easily set up or taken down, making hot water portable and affordable, even for apartment dwellers.

Before we take a closer look at the types of spas and hot tubs available today, let's examine a few basics. First, we offer a few definitions:

FIGURE 1-1 Japanese snow monkeys in natural hot springs.

FIGURE 1-2 Natural hot springs in Yellowstone National Park.

- *Spa* generally refers to a small pool of up to 3000 gallons (11.4 cubic meters) of water that is agitated by jet action and/or the introduction of bubbles into the water. The spa is built either in the ground, just as any other concrete and plaster in-ground swimming pool (Fig. 1-3), or above ground as a self-contained unit consisting of a fiberglass tub, served by a small package of pumps, motors, air blowers, filters, and controls (Fig. 1-4).

- *Hot tub* generally refers to a wooden vessel made of cedar, redwood, or other natural materials (Fig. 1-5) that is filled with water and serviced by pumps, motors, air blowers, heaters, and other standard spa equipment (sized appropriately for the smaller volumes of water). Hot tubs can be built for 10 or more adults, but are typically designed for four adults and contain no more than 1500 gallons (5.7 cubic meters) of water.

- *Portable spa* can refer to any single unit that contains the vessel and all related equipment and that is easily moved from one location to another (Figs. 1-6 and 1-8). That said, *portable* today means the latest lightweight and soft-sided units that can be easily folded up and stored in a space not much bigger than the trunk of your car. That's truly portable!

For simplicity, I refer to *spa* throughout the book, meaning either a spa or a hot tub, unless we're dealing with a subject that is unique to hot tubs themselves. Next, let's understand how a spa works.

How It Works

Figure 1-6 shows the layout of a typical spa and related equipment. For ease of understanding, this drawing shows the equipment and the

A   C

FIGURE 1-3 **Typical in-ground spas.** *B: Medallion Spas. C: Bradford Spas.*

vessel in one wooden structure, but the equipment can also be located some distance from the body of water itself. Either way, the plumbing and concepts will be similar.

To understand how a spa works, just follow the path of the water, starting in the tub itself. That is also how this book will be outlined—in the logical pattern that the water travels from spa through plumbing to the pump/motor, filter, and heater (and/or solar heating system) and back to the spa.

The water enters the plumbing through a main drain and/or a surface skimmer (Fig. 1-6, item 7). Most spas have both, while hot tubs typically have only a main drain. Most equipment packages direct the water from the pump to the filter, but portable spas with built-in filters pass the water through the filter (item 8) on the way to the the circulation pump (item 14). From there the water is warmed by a heater

A

B

FIGURE 1-4 **Typical above-ground spas.** *A: Nordic Hot Tubs. B: Bradford Spas.*

FIGURE 1-5 **Typical redwood hot tub.** *Gordon & Grant Hot Tubs.*

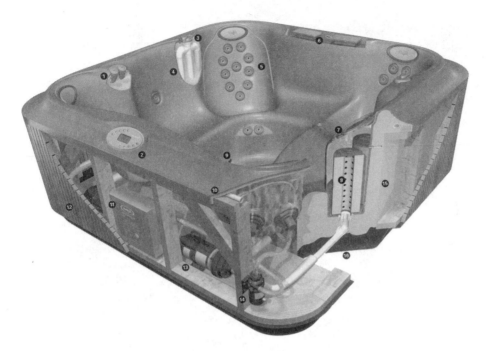

1 Air button equipment controls
2 Electronic equipment controls
3 Jet flow diverter valves
4 Jet plumbing
5 Massage jets
6 Waterfall return flow
7 Skimmer grill
8 Built-in cartridge filter system
9 Spa drain
10 Acrylic coating
11 Electronic control and wiring junction box
12 Redwood-style cabinetry
13 Jet pump
14 Circulation pump
15 Insulation
16 ABS foundation material

FIGURE 1-6 **Layout of typical spa.** *Jacuzzi Premium.*

(inside item 11), then returned to the spa through return ports (item 6). This return flow is typically pumped with sufficient force to provide a strong jet action, which gives spas and hot tubs their distinctive therapeutic massage characteristics. In some spas, a separate jet pump (item 13) provides the power to a series of dedicated jets (item 5).

How Much Water Does It Hold?

Many of the procedures described in this book, especially those involving chemicals and maintenance, require knowing how much water is in your spa or hot tub. These calculations are particularly important in small bodies of water, where improper estimates could lead to serious errors in the chemical balance of the water and possible injury to those using the spa.

Many spas will have information provided with them, sometimes on a nameplate affixed to the vessel itself, to tell you exactly how many gallons or liters of water are in your unit. If the labels or owner's manuals are missing, or if you choose not to fill the vessel to its capacity, you will need to know how to calculate the volume of water for yourself.

Square or Rectangular

The formula is simple:

Length × width × average depth × 7.5 = volume (in gallons)

Let's first examine the parts of the formula. Length times width gives the surface area of the spa. Multiplying that by the average depth determines the volume in cubic feet. Since there are 7.5 gallons in each cubic foot, multiply the cubic feet of the spa by 7.5 to arrive at the volume of the spa (expressed in gallons).

The formula is simple and so is the procedure. Measure the length, width, and average depth of the spa, rounding each measurement to the nearest foot or percentage of 1 foot. If math was not your strong suit in school, each inch equals 0.0833% of 1 foot. Therefore, multiply the number of inches in your measurements by 0.0833 to give you the appropriate percentage of 1 foot when the measurement does not equal exactly a certain number of feet. For example,

$$9 \text{ feet, 9 inches} = 9 \text{ feet} + (9 \text{ inches} \times 0.0833)$$

$$= 9 + 0.75$$

$$= 9.75 \text{ feet}$$

Average depth will be only an estimate, especially in spas with irregular seats and contours for sitting or lying down. Figure 1-7A shows at least two different depths. You will need to estimate how much of the spa is at one depth and how much at the other. If, for example, you estimate that the spa is equally divided between seat area and standing area and that the depth of the seat portion of the spa is 2 feet, while the center section of the spa is 5 feet, then the average

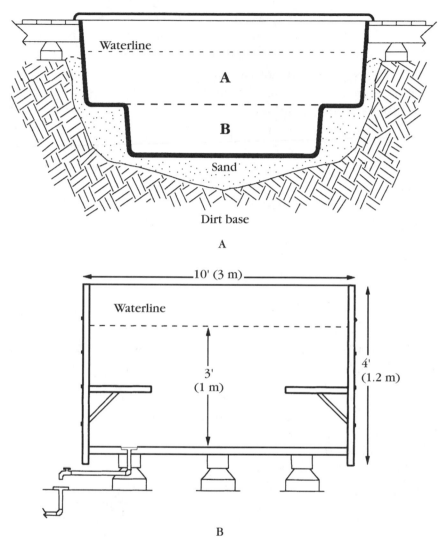

FIGURE 1-7 (A) Cutaway view of typical spa. (B) Cutaway view of typical hot tub.

of those two sections is 3.5 feet. Use that average for your depth figure when calculating the volume.

But if most of the spa is one depth and smaller parts are another depth, you may want to treat it as two parts. Measure the length, width, and average depth of the shallow section; then make the same measurements for the deeper section. Next calculate the volume of the shallow section, and add that to the volume you calculate for the deeper section.

In either case, be sure to use the actual water depth in your calculations, not the depth of the container. For example, the hot tub depicted in Fig. 1-7B is 4 feet deep, but the water will only be filled to about 3 feet. Using 4 feet in this calculation would result in a volume 33% greater than the actual amount of water. This could mean serious errors when adding chemicals, for example, which are administered based on the volume of water in question. There might be a time when you want to know the potential volume, if filled to the brim. Then, of course, you would use the actual depth (or average depth) measurement. In the example, that was 4 feet.

Try to calculate the volume of a spa that measures 10 feet by 9.75 feet and is 4 feet deep:

$$\text{Length} \times \text{width} \times \text{average depth} \times 7.5 = \text{volume (in gallons)}$$

$$9.75 \text{ feet} \times 10 \text{ feet} \times 4 \text{ feet} \times 7.5 = 2925 \text{ gallons}$$

The same formula works in the metric system. Since metric units are already based on a decimal system in units of 10, there is no need to convert anything when you are dealing with fractions of a meter or other units of measurement. As with standard units, round to the nearest decimal for ease of calculation:

$$\text{Length} \times \text{width} \times \text{average depth} = \text{cubic meters of volume}$$

That's as far as you will need to take the equation, since things such as spa chemical dosages in metric measurements will be based on a certain amount per cubic meter. Dosages may also be expressed in certain amounts per liter of water. Since there are 1000 liters in 1 cubic meter, the formula becomes

$$\text{Length} \times \text{width} \times \text{average depth} \times 1000 = \text{volume (in liters)}$$

Circular

The formula:

$$3.14 \times \text{radius squared} \times \text{average depth} \times 7.5 = \text{volume (in gallons)}$$

The first part, 3.14, refers to pi, which is a mathematical constant. It doesn't matter why it is 3.14 (actually the exact value of pi cannot be calculated). For our purposes, we need only accept this as fact.

The radius is one-half the diameter, so measure the distance across the broadest part of the circle and divide that in half to arrive at the radius. Squared means multiplied by itself, so multiply the radius by itself. For example, if you measure the radius as 5 feet, multiply 5 feet by 5 feet to arrive at 25 feet. The rest of the equation was explained in the square or rectangular calculation.

Use the spa in Fig. 1-7B to calculate the volume of a round container. Let's do the tricky part first. The diameter of the spa is 10 feet. One-half of that is 5 feet. Squared (multiplied by itself) means 5 feet times 5 feet = 25 square feet. Knowing this, we return to the formula:

$$3.14 \times \text{radius squared} \times \text{average depth} \times 7.5 = \text{volume (in gallons)}$$

$$3.14 \times 25 \text{ feet} \times 4 \text{ feet} \times 7.5 = 2355 \text{ gallons}$$

Once again, when measuring the capacity of a circular spa, you may need to calculate two or three areas within the spa and add them to arrive at a total volume. An empty circular spa looks like an upside-down wedding cake, because of the seats, as in Fig. 1-7A. Therefore, you might want to treat that as two separate volumes—the volume above the seat line and the volume below. In the wooden hot tub depicted in Fig. 1-7B, where there is actually water above and below the seat, the tub can be measured as if there were no seats, since this difference is negligible.

The calculations in metric units will be the same, but remember that there is no need to multiply by 7.5 to determine the volume in cubic meters. To learn the volume in liters, however, multiply by 1000 instead of 7.5, and the result will be expressed in liters of water.

How Is It Made?

Virtually every waterproof material is now used to make spas, creating unique designs that are lighter, more durable, and more attractive than ever before. Let's take a look at the astonishing array of styles and materials used to make spas and hot tubs today.

Concrete and Plaster

Look back at Fig. 1-3, which shows typical inground concrete spa and pool combinations, both finished with fine plaster. Many homes today feature a concrete and plaster spa by itself as busy families may find greater use for a spa than for a pool. Spas take up considerably less space, are less costly to build and operate, and require far less maintenance than pools. Pumps, filters, heaters, and other equipment are usually located some distance from these spas, to keep noise away from spa users. Plumbing is typically buried underground, just as it would be for an inground pool.

Fiberglass and Acrylic

Spas made of fiberglass are hand-crafted to conform to a specific design requirement. Sheets of fiberglass material provide the strength, while chemical resins coat the material and provide stiffness and strength. The final shell is covered with a glossy coating, called the *gel coat,* to provide a surface that is virtually impervious to water, heat, or harsh chemicals.

Acrylic spas are essentially big molded-plastic bowls, and they can be treated much the same as fiberglass spas. Both fiberglass and acrylic spas are built as shells that can be plumbed and set into the ground, mounted above ground, or built into a self-contained cabinet for portability (Fig. 1-8). The self-contained unit requires only the addition of water and a 30-amp circuit (the largest units may require heavier electrical supply and wiring) to plug in the equipment. Although they are portable, the weight of the shell, cabinet, and equipment package makes these self-contained units at least as difficult to move as a grand piano.

Soft-Sided

As an alternative to the weight of that grand piano, several manufacturers have developed a lightweight, very portable, self-contained spa made from reinforced foam (Fig. 1-9). As the name implies, the soft-sided spa is made of stiff, but flexible, foam rubber, reinforced with aluminum framing and covered with a waterproof fabric to hold water. These spas are built with a self-contained equipment package, usually covered in the same materials to protect the equipment and reduce noise. Many apartment dwellers are buying these spas because of their true portability and low cost.

Wood

As we have seen in Fig. 1-4, fiberglass or acrylic spas are often finished with a wooden exterior for both aesthetic and insulation purposes. But

A

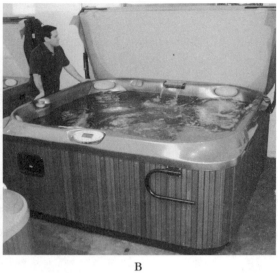

B

C

FIGURE 1-8 **Typical self-contained fiberglass spas.** *A: Master Spas. C: Sundance Spas.*

FIGURE 1-9 Typical soft-sided spa.

many people prefer a hot tub made entirely of natural wood, using simple designs and practices that have been used for centuries to store wine and to build wooden boats.

Figure 1-5 showed the typical redwood hot tub, still as popular today as when the ancient Greeks first built wooden baths by joining planks together. Wooden hot tubs add a natural beauty to backyard gardens, but require significantly greater maintenance and are far less portable than spas made from artificial materials. If properly maintained, however, wooden hot tubs can last for 20 years or more.

Aluminum Shell, Vinyl-Lined

Spas made like above-ground pools are growing in popularity because of their portability and affordability. Figure 1-10 shows a spa made with an aluminum shell, then made waterproof with a vinyl liner. The spa is finished with a cabinet of wood or manufactured materials.

FIGURE 1-10 Typical aluminum-shell, vinyl-lined spa. *Medallion Spas.*

Stainless Steel

Few consumers will pay the high cost of stainless steel when buying a spa, but for those who want exceptional durability and style, this may be just the right choice. Figure 1-11 shows a custom-built inground spa, made entirely of stainless steel. Figure 1-12 shows a more portable version with ceramic tile accents. Anytime that metal comes into contact with water, heat, and chemicals, corrosion is likely, unless extra care is taken with maintenance and water chemistry. With that said, stainless steel spas are both functional and quite handsome.

Jetted Bathtubs

Another choice of spa, especially in homes or apartments where space is an issue, is the jetted bathtub. As the name implies, a standard bathtub is fitted with a booster pump and jets to create a spa (Fig. 1-13). Some units are built to accommodate two people or built round instead of in the traditional rectangle. Jetted tubs are filled as any other bathtub with hot water from the normal home water heater supply, so no additional heater is used.

FIGURE 1-11 Custom stainless steel spa. *Diamond Spas.*

Since they are usually part of the indoor bathroom and are drained after each use, jetted tubs do not have skimmers, strainer baskets, or filters. Jetted tubs are popular as "personal" spas since they are inexpensive, requiring only one pump and some plumbing; they heat up quickly, since they are filled with water that is already hot; and they require no water maintenance or special chemicals. Obviously the jetted tub is maintained as any other bathtub, drained, and cleaned after each use.

The Rest

Virtually any container that holds water and is large enough to accommodate bathers can be made into a spa (Fig. 1-13D). In various parts of the world, I have seen spas made from large metal or plastic barrels, old water tanks and bathtubs, as well as the more traditional construction types.

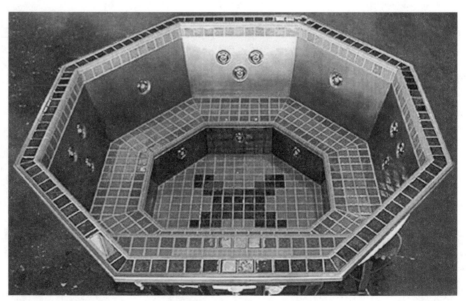

FIGURE 1-12 Stainless steel spa with tile. *Bradford Spas.*

A

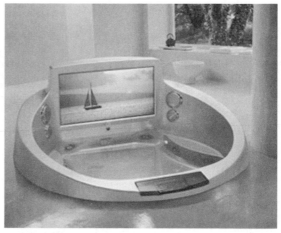

B

C

D

FIGURE 1-13 **(A) Typical jetted bathtub. (B) Jetted bathtub with entertainment center. (C, D) Custom bath spa.** *B: Jacuzzi Premium. C, D: Diamond Spas.*

New materials and designs, especially using new plastics and composites, are added every year, but a general understanding of the majority of styles and materials will help you make a selection, no matter what you find locally. Typically, manufacturers are constantly striving to add models that are lighter and cheaper, bringing the pleasures of spa ownership to more people every day.

Selecting a Spa or Hot Tub

Now that we have examined the types of spas and hot tubs generally available, let's consider some of the criteria you might use to make a selection that's right for your personal or family use. Actual installation procedures are explained in later chapters, but these guidelines will help you make your spa investment. For the purposes of this section, the *shell* refers to the actual vessel, regardless of the material used in its fabrication.

Location

Where do you plan to put your spa? If it will be set among the trees and garden, you might prefer a more natural look, choosing a wood tub or a wood cabinet for an acrylic spa (Fig. 1-14). If the spa will be located on a deck, can the weight of the unit and the water be safely supported by that structure?

The weather also plays a factor in determining your selection. For example, in colder climates you might prefer an inground spa for greater insulation and protection against frozen plumbing. In windy areas, you will want to be sure your spa can be covered to prevent heavy debris from fouling the water or excess evaporation. Since the hot water will be already heating the bathers, you may want to select a spa that can be located under some type of shade, from your house, a separate awning, or trees (Fig. 1-15).

If you want your spa indoors or on a small patio, such as in an apartment, you will want to consider the limited space when selecting your spa. If you think you might move often (or just want to move the spa frequently), consider one of the more portable models.

The Shell

What is the maximum number of people who will use the spa at any given time? Spas are sold as two-person, four-person, etc., indicating

FIGURE 1-14 Shaded spa placement. *Great Northern Hot Tubs.*

FIGURE 1-15 Yard placement of spa. *Master Spas.*

the number of bodies that the shell can comfortably accommodate. If you are considering a model with molded seating or lounges, the capacity will be fairly obvious. (See Fig. 1-16.)

Don't forget to consider overall use of the spa as well as your day-to-day use. For example, you may have one or two parties each summer when 8 or 10 people would use the spa at a time, but the remainder of the year it will only be used by 2 people at a time. The amount of space the 10-person spa requires on the deck or in the yard and the annual

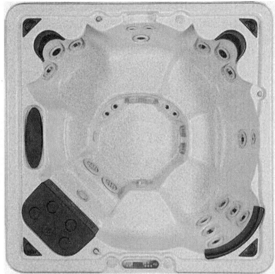

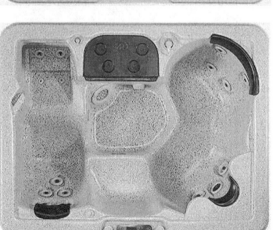

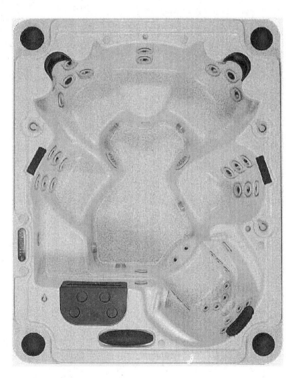

FIGURE 1-16   Spa shell varieties. *Master Spas.*

costs of maintenance, heating, and electricity may not make sense if it is used mostly by only two people.

Consider the manner in which the spa will mostly be used. You will want straight bench seating if it is used mostly for socializing with several people, but will prefer lounges if it is used for longer, therapeutic soaking.

Also make note of the size of the most frequent users. I was once asked to install a spa for a very famous actor/musician in southern California who is also very short. His builder had already purchased the shell, preplumbed with skimmer, drains, and jets. When the project was complete, this person sat in his new spa and found the water level up to his eyeballs. The seating had been designed for a taller person whose upper body length would leave the shoulders at the waterline. Most spas have seats of varying depth to accommodate a variety of human proportions, but the choice of this one had not taken into account that its only occupant was so short! The solution was to lower the skimmer so we could also lower the waterline, something we all should have thought about first.

The Plumbing

Similar factors need to be considered when plumbing the shell. Depending on the intended users, you may be able to install a shell that has been preplumbed to accommodate the average bather. Typically, two jets per person are included in a spa, set in the walls at varying heights. An air bubble ring is also provided to create general turbulence in the water for the massaging effect.

If you have specific preferences about the number, type, and location of the jets, it makes more sense to buy the shell and plumb it to those specifications. By inquiring at your spa retailer or looking at selections online, you can make a list of available options such as small "pinpoint" jets, larger standard massage jets, rotating jets, and those that can be turned on or off to put more or less pressure in one specific area of the spa. (See Fig. 1-17.)

Some people care about only the general turbulence of hot water and require one or more air bubble rings, while others are interested in only the massage jets and don't want to invest the extra money in the air bubble ring and electric blower required to power it. A blower can also be plumbed to "turbocharge" the jets. In short, carefully think through your preferences, and find the spa that actually meets those needs.

FIGURE 1-17 Jet options.

When you are selecting a preplumbed spa, consider the cost and feasibility of transportation. Figure 1-18 shows a pre-plumbed spa shell. As you can see, the exposed fittings and plumbing can easily be cracked or snapped off during transport and handling to the job site. I learned my lesson early when I failed to personally supervise the delivery of a preplumbed spa and simply installed it for a customer. After many hours of excavation to expose the plumbing all around, I found four leaks—cracks that were probably caused by careless handling during transportation. There is an irresistible urge to use the plumbing fittings as handles, which invariably weakens or outright cracks them.

The moral of the story is to include a truck of adequate size and enough handlers (remember, these things are large, bulky, and *heavy*) to get the spa into the yard without damage. If the pathways are tight or the location is far from the truck access area, plumb the spa at the

FIGURE 1-18 Preplumbed spa shell. *Bradford Spas.*

installation site. Of course, if you are selecting a complete packaged spa in a self-contained cabinet, it is less likely to be damaged, but is even more bulky than the shell by itself. Once again, consider the ultimate location of your spa, and measure the doorways, gates, and pathways from the street to the job site.

The Cost

Finally, price makes a difference to almost everyone. The self-contained spa, built with the same shell and plumbing you might find in a custom design, will cost anywhere from $2000 to $8000. By the time you factor in the installation of the shell, plumbing, equipment, and subcontractors, you can easily spend up to 10 times that much for a custom-built spa.

Even if you choose a prepackaged spa that includes all the equipment, there are numerous upgrades and options for any spa, which will further determine the final cost of the project. At the same time as you are choosing the number and styles of jets and plumbing options, you need to review the equipment requirements of each choice.

The pump needs to be of sufficient strength to provide adequate power to the jets. The general rule of thumb is that each jet needs ¼ horsepower, so a 2-horsepower pump could theoretically give adequate performance to eight jets. Given that, however, you must also consider the hydraulics of the system as described in Chap. 3, "Pumps and Motors," especially if the equipment is located a great distance from the spa. Long distances are often required to ensure that the equipment noise is kept away from the spa environment.

Using those general guidelines, determine the size of pump needed and whether more than one pump is required. In larger installations, you may want to divide the spa into two halves and operate each side from a separate pump. When making pump selections, consult your electrician to determine that adequate electricity is available for your choices.

Cartridge filters are the preferred choice for spas, following selection and sizing criteria described in Chap. 4, "Filters." Consider the volume of use when choosing a filter for your spa. If you plan to use the spa frequently (many people jump in their spas when they get home from work before greeting the rest of the family!), choose a filter size larger than actually needed to reduce the frequency of required teardown and cleaning. If you plan to hire a pool technician to maintain your spa, you might be less concerned about the frequency of filter cleaning; but most do-it-yourselfers rarely clean the filter as often as needed.

One of the most important features of a spa is the heat in the water itself. The most important consideration in choosing a heater is, once again, your intended use. If your use is unpredictable, you will need a larger heater to be able to raise the temperature quickly. On the other hand, if you know the spa needs to be heated at the same time each day, you can choose a smaller heater and set the system to turn on with a time clock early enough to reach the desired temperature by the desired time.

Most small spas are heated with an electric heater, but larger heaters (and those with the fastest heating time) are gas-fueled. Consult with your plumber about the availability of gas supply, and consider the cost of upgrading lines, meters, and adding the supply line to the equipment location. If there is no gas available in your area, consult with your electrician about adequate electrical supply for the electric heater. Since even the largest electric heaters take a long time to heat a spa, choose the largest one you and the electrical supply can afford.

TRICKS OF THE TRADE: ADDING SPA ACCESSORIES

To provide a true overall estimate of the job, consider this list of additional equipment you might want to add to your new spa or hot tub.

- Cover
- Chlorinator
- Lighting
- Mist spray
- Fill line (automatic or manual)
- Control system (electronic, manual, or air buttons)
- Time clock (may be included in control system)
- Handrails, built-in drink holders, padded neck rests

Your local spa retailer, or a variety of websites, can provide catalogs of available products to help you make these choices; but don't forget to add the cost of them to arrive at a realistic budget for your project.

Decide whether you want one or more blower rings, which force bubbles into the water for a general massage. Rings are often located in the floor and/or seats of the spa. You might add a second blower to turbo-charge the jets, or install a three-port valve to divert the airflow of a single blower between the air ring and jets. Again, consult with the electrician about available electrical supply and cost of wiring to the equipment location for booster pumps or blowers that service blower rings.

Plumbing Systems

Now that we have examined the types of spas and hot tubs and you have selected the one that's right for you, we will take a look at the network of pipes, valves, and jets that circulate the water. Although your spa may be preplumbed and all of these components hidden from view, you may want to repair or upgrade your spa someday, so a basic understanding of the plumbing systems is essential.

To truly understand a spa, we must follow the path of the water. As can be seen from Fig. 2-1, water from the spa enters the equipment system through a main drain on the floor, through a surface skimmer, or through a combination of both main drain and skimmer. The water travels into a pump that is driven by the attached motor. From the pump, the water travels through a filter, then to the heater, and finally back to the spa through jets that provide both clean water and the characteristic massage action that makes spas so popular in the first place.

Skimmers

The purpose of the skimmer, as the name implies, is to pull water into the system at the surface with a skimming action, pulling in leaves, oil, and dirt before they can sink to the bottom of the spa. The skimmer also provides a conveniently located suction line for vacuuming the

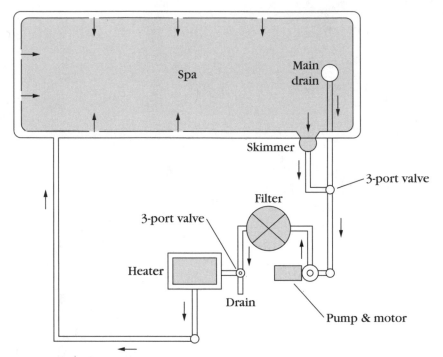

FIGURE 2-1 Plumbing diagram of spa.

spa. Most skimmers today are molded, one-piece plastic units. Older inground spas have built-in-place concrete skimmers.

In either case, the skimmer will be accessed through the face of the spa wall (Fig. 2-2) or through a cover that sits on the deck at the edge of the spa (the cover will be plastic or concrete). Most redwood hot tubs have only a main drain; but if they also include a skimmer, it is likely to be a flat, vertical model that has no basket but skims the surface and pulls any floating debris to a plastic screen (Fig. 2-3). Modern portable spas, with filter cartridges built right into the spa, incorporate a skimmer directly above these cartridges (Fig. 2-4).

So how does the typical skimmer work? Water pours over a floating weir (Fig. 2-2) which allows debris to enter; but when the pump is shut off and the suction stops, the weir floats into a vertical position and prevents debris from floating back out to the spa. Debris can cause the weir to jam in a fixed position, thus preventing water from flowing into the skimmer. When this happens, the pump will lose prime (water flow) and run dry, causing damage to its components. Therefore, during windy periods it may be better to remove the weir from the skimmer to prevent such problems.

Spa skimmers also include a debris basket, which will collect leaves and large debris and is easily emptied. The hot tub skimmer (Fig. 2-3) pulls the debris to the grid, but when the suction stops, the debris is free to float back into the tub. The skimmer built into the portable spa (Fig. 2-4) includes a debris bag, much like the ones on vacuum cleaners, which can be emptied periodically.

By the way, you should exercise care in working around the skimmer when the pump is on. I have nearly had fingers broken when placing my hand over a skimmer suction opening and have lost various pieces of equipment, T-shirts, bolts, plastic parts, etc. which invariably end up clogged in the pipe at some turning point where leaves, hair, and debris later catch and close off the pipe completely. Keep small objects away from the skimmer opening when the basket is removed, and especially keep your hands from covering that suction hole!

Main Drains

If your spa also has a main drain, it will be connected to the skimmer, and then both units are connected to the pump. A three-port valve is used to adjust the amount of suction taken from the main drain and the amount taken from the skimmer.

A

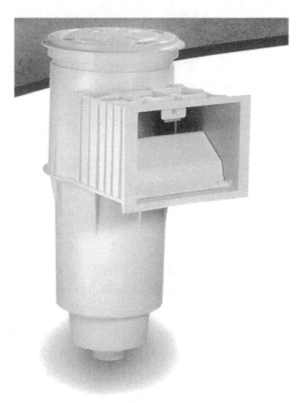

B

FIGURE 2-2 Spa skimmer: (A) front access style in use; (B) front access–style hardware.

FIGURE 2-3 Hot tub skimmer.

FIGURE 2-4 Spa skimmer—top access.

In some spas, there may be more than one main drain, so that if one becomes covered with a foot or hand, water will be pulled from the other, avoiding injury to the bather. These drains are located at least 12 inches (30 centimeters) apart. Because spas are relatively shallow, strong suction can create a whirlpool effect. To prevent this, many spa main drains are fitted with antivortex drain covers, which are slightly domed-shaped with the openings located around the sides of the dome (Fig. 2-5).

Equipment alone can't prevent accidents. Children playing around the main drain or even casually placing feet or hands over the suction can be injured. Long hair can easily become trapped. The best way to avoid such accidents is to be sure that anyone using the spa has been given adequate training and warnings; that appropriate warning signs are posted; and that the safety covers and other equipment designed to prevent accidents are maintained at all times.

General Plumbing Guidelines

RATING: ADVANCED

Most spa plumbing will consist of common PVC (polyvinyl chloride) plastic, either rigid or flexible varieties, in 1½- or 2-inch (40- or 50-millimeter) diameter, referring to the interior diameter (the diameter of the pipe that is in actual contact with the water). Before we proceed to specific instructions on working with PVC, here are a few useful general guidelines.

Measure the pipe run carefully, particularly if you are repairing a section between plumbing that is already in place. In measuring, remember to include the amount of pipe that fits inside the connection fitting, usually about 1½ inches (40 millimeters) at each joint (Fig. 2-6).

When you are working on in-place plumbing, support your work by building up wood or bricks under the pipe on each side of your work area. This will prevent vibration as you cut, which can damage pipe or joints farther down the line. Also, unsupported pipe will sag and bind when you cut it. That is, as you cut, it pinches the saw blade, making cutting difficult—and straight, clean cuts impossible.

FIGURE 2-5 Main drain and antivortex cover.

Threaded fittings are obvious and simple; however, leaks occur most often in these connections. The key is to carefully cover the male threads with Teflon tape and to tighten the fitting as far as possible without cracking.

Teflon tape "fills" the gaps between the threads to prevent leaking. Apply the tape over each thread twice, pulling the tape tight as you go, so that you can see the threads. Apply the tape clockwise (Fig. 2-7) as you face the open end of the male threaded fitting. If you apply the tape

FIGURE 2-6 Measuring pipe and fittings.

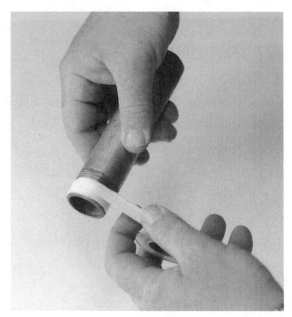

FIGURE 2-7 Correct application of Teflon tape.

backward, when you screw on the female fitting, the tape will "skid" off the joint. Try it both ways to see what I mean, and you will make that mistake only once!

Another method of sealing threads is to apply joint stick or pipe dope. These are odd names for useful products that are applied in similar ways (Fig. 2-8). *Joint stick* is a crayon-type stick of a gumlike substance that works like Teflon tape. Rub the joint stick over the threads so that the gum fills the threads. Apply *pipe dope* the same way. The only difference is that pipe dope comes in a can with a brush and is slightly more fluid than joint stick. The key to success with joint stick or pipe dope is to apply it liberally and around all sides of the male threaded fitting, so that you have even coverage when you finally screw the fittings together. Some product will ooze out as you tighten the fittings, but that proves that you have applied enough.

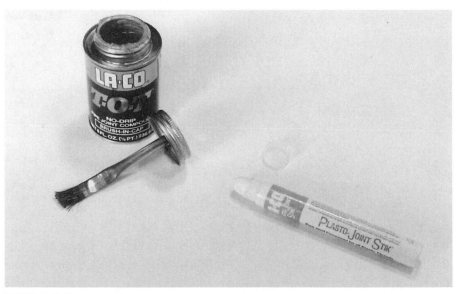

FIGURE 2-8 Pipe dope and joint stick.

I use Teflon tape because I know upon application that it is an even and complete coverage of the threads. Pipe dope or joint stick may not apply evenly or may bunch up when you are threading the joint together. If you do use dope or stick, be sure it is a non-petroleum-based material (such as silicone). Petroleum-based products will dissolve plastic over time, creating leaks.

When you are working with PVC pipe and fittings, tighten threaded fittings with a *channel lock* type of pliers of adequate size to grip the pipe. Using pipe wrenches usually results in application of too much force and cracking of the fittings. If you don't have pliers large enough for the work and must use pipe wrenches, tighten the work slowly and gently.

PVC Plumbing

Spa plumbing is generally made with lengths of PVC plastic pipe and connection fittings of the same materials that join those lengths. The pipe acts as the "male" which is glued into the "female" openings of the connection fittings. Some connection fittings are threaded for assembly that is even easier.

PVC is manufactured in a variety of different strengths depending on the intended use. To help identify the relative strength of PVC, it is labeled by a *schedule* number; the higher the number, the heavier and stronger the pipe. Spa plumbing is done with PVC Schedule 40. Some gas lines are plumbed with PVC Schedule 80.

Ultraviolet light from the sun will cause PVC to become brittle over time, losing strength under pressure and creating cracks. Chemical inhibitors are added to some PVC to prevent this, the most common and cheapest being simple carbon black (which is why plastic pumps and other pool/spa equipment is often made black). Another common preventive measure is to simply paint any pipe that is regularly exposed to sunlight.

PVC pipe also is manufactured in a flexible variety, making plumbing easier around tight spaces around spas and jetted tubs (Fig. 2-9). Flex PVC is available in colors for cosmetic purposes and has the same characteristics and specifications as rigid PVC of the same schedule and size.

PVC pipe is connected with fittings (Fig. 2-10). Fittings allow connection of pipe along a straight run (called *couplings*), right angles

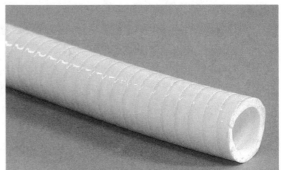

FIGURE 2-9   Flex PVC.

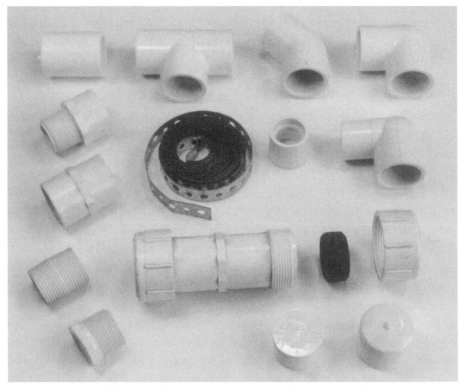

FIGURE 2-10   PVC fittings.

(called *90-degree couplings* or *elbows*), 45 degree angles, T fittings, and a variety of other formats. Fittings that are smooth-fitted and glued together are called *slip fittings,* while those with threads are called *threaded fittings.*

In most spa plumbing, the long runs of pipe will be underground. Sometimes, however, horizontal runs will be under a house or deck or over a slope where support is needed. In this case, pipe should be supported every 6 to 8 feet (2 to 2.5 meters), hung with plumber's tape to joists, or supported with wooden bracing. PVC does not require support on vertical runs because of its stiffness, but common sense and local building codes may require strapping it to walls or vertical beams to keep it from shifting or falling over. Remember, the pipe becomes considerably heavier when filled with water and may vibrate along with pump vibration.

Plumbing Methods

RATING: ADVANCED

The concept of joining PVC pipe involves welding the material together by using glue that actually melts the plastic parts to each other. In truth, each joint will have an area that is slightly tighter than the rest. In the tightest parts, this welding actually occurs. In the remainder, the glue bonds to each surface and itself becomes the bonding agent. Obviously the strongest part of each joint is the welded portion; but in either case, the key is to use enough glue to ensure total coverage of the surfaces to be joined.

Following is the correct procedure for plumbing with PVC (Fig. 2-11):

1. **Cut and Fit** Cut and dry-fit all joints and plumbing planned. It is easy to make mistakes in measuring or cutting, and sometimes fittings are not uniform, so they don't fit well. Dry fitting ensures the job is right before gluing. If you need the fitting and pipe to line up exactly for alignment with other parts, make a line on the fitting and pipe (Fig.

TOOLS OF THE TRADE: PVC PLUMBING

The supplies and tools you need for PVC plumbing are

- **Hacksaw with spare blades (coarse: 12 to 18 teeth per inch or 2.5 centimeters)**
- **PVC glue and primer**
- **Cleanup rags**
- **Fine sandpaper**
- **Teflon tape or joint stick**
- **Waterproof marker**

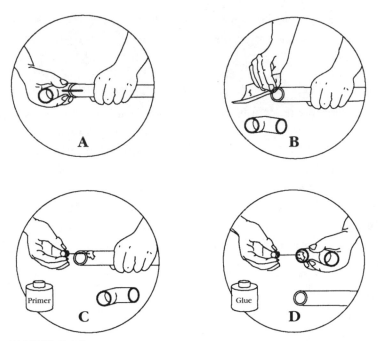

FIGURE 2-11 Step-by-step PVC plumbing.

2-11A) with a marker when dry fitting so you have a reference when you glue them together.

2. **Sand** Lightly sand the pipe (Fig. 2-11B) and inside the fittings so they are free of burrs. The slightly rough surface will also help the glue adhere better.

3. **Prime** You might need to apply a preparation material, called *primer*, to the areas to be joined before gluing (Fig. 2-11C). Some PVC glues are solvent/glue combinations, and no primer is required. In some states, however, use of primer might be required by the building code, so check that before selecting an all-in-one product. If you are using primer, apply it, with the swab provided, to both the pipe and the inside of the fitting. Read and follow the directions on the can.

4. **Glue** Before gluing, be ready to fit the components together quickly because PVC glue sets up in 5 to 10 seconds. Apply glue to the pipe and to the inside of the fitting (Fig. 2-11D).

5. **Join** Fit the pipe and fitting together, duplicating your dry fit, and twist about a half turn to help distribute the glue evenly, realigning the

TRICKS OF THE TRADE: PVC PLUMBING

1. Make all threaded connections first, so if you crack one while tightening, it can be easily removed. Then glue the remaining joints to the threaded work.

2. When cutting PVC pipe, hacksaw blades of 12 teeth per inch (per 2.5 centimeters) are best, particularly if the pipe is wet (as when making an on-site repair). Finer blades will clog with soggy, plastic particles and stop cutting. Use blades of 10 inches (25 centimeters) in length. They wobble less than 12- or 18-inch blades during cutting. In all cases, the key is a fresh, sharp blade. For the few pennies involved, change blades in your saw frequently rather than hacking away with dull blades—you'll notice the difference immediately.

3. No matter how careful you are, you will drip some glue on the area or on yourself. That's why I always carry a supply of dry, clean rags to keep myself, the work area, and the customer's equipment clean of glue.

4. Try to make as many *free* joints as possible first. By that I mean the joints that do not require an exact angle or that are not attached to equipment or existing plumbing. The free joints are those that you can easily redo if you make a mistake. Do the hard ones last—those that commit your work to the equipment or existing plumbing and cannot be undone without cutting out the entire thing and starting over.

5. Use as much glue as you need to be sure there is enough in the joint. It's easier to wipe off excess glue than to discover that a small portion of the joint has no glue and leaks.

6. Practice. PVC pipe and fittings are relatively cheap, so make several practice joints and test them for leaks in the shop before working on someone's equipment in tight quarters in the field.

7. Flexible PVC is the same as rigid, but when you insert the pipe into a fitting, hold it in place for a minute or longer because flex PVC has a habit of backing out somewhat, causing leaks.

8. In cold weather, more time is required to obtain a pressure-tight joint, so be patient and hold each joint together longer before going on to the next.

9. Bring extra fittings and pipe to each job site. Bring extras of the types you expect to use, as well as types you don't expect to use, because you just might need them. Nothing is worse than completing a difficult plumbing job and being short just one fitting, or needing to cut out some of your work and not having the fittings or a few feet of pipe to replace them. It is often several miles back to the office or the nearest hardware store to grab that extra fitting that should have been in your truck in the first place. Bring extra glue, sandpaper, and rags, too.

lines drawn on the pipe and on the fitting. If you are using flexible PVC, because it is made by coiling a thin piece of material and bonding it together, do not twist it clockwise. This can make the material swell and push the pipe out of the fitting. Get in the habit of twisting all pipe *counterclockwise* (even though it makes no difference with rigid PVC), and you will never make that mistake.

6. **Seal** With rigid PVC, hold the joint together about 1 minute to ensure a tight fit; about 2 minutes with flex PVC. Although the joint will hold the required working pressure in a few minutes (and long before the glue is totally dry), allow overnight drying before running water through the pipe to be sure. I have seen demonstrations with some products (notably Pool-Tite solvent/glue) where the gluing was done underwater, put immediately under pressure, and held just fine. I don't, however, recommend this procedure as I have gone back on too many plumbing jobs to fix leaks a few weeks later because I hastily fired up the system after allowing only a few minutes of drying time.

Manual Three-Port Valves

The design of three-port valves takes water flow from one direction and divides it into a choice of two other directions.

Picture a Y, for example, with the water coming up the stem, then a diverter allows a choice between one of two directions (or a combination thereof). Conversely, the water flow might be coming from the top of the Y, from two different sources, where the diverter decides which source will continue down the stem, or mix some from each together.

Figure 2-12 shows a typical Y or three-port valve. A valve body (item 7), built to 1½- or 2-inch (40- or 50-millimeter) plumbing size, houses a diverter (item 8) which is moved by a handle (item 3) on top of the unit. Typically, these valves are used when a pool and spa are both operated from the same pump/filter/heater equipment. As noted previously, these valves can also divide the suction between a main drain and a skimmer. As the water returns to the spa, a three-port valve can divide water between various jets.

Construction

The unit is made watertight by gaskets (items 5 and 6), and the diverter is held in the valve body by a cover (item 4) which secures to the housing with metal screws (item 1).

Maintenance and Repair

As the simple parts suggest, there's not too much that can go wrong with manual three-port valves; however, a few procedures are common.

INSTALLATION

RATING: ADVANCED

Most three-port valves are glued directly to PVC pipe, using regular PVC cement as in any other fitting. Care should be exercised not to use too much glue, as excess glue can spill onto the diverter and cement it to the housing. Excess glue can also dry hard and sharp, cutting into the gasket each time the diverter is turned, creating leaks from one side to the other.

LUBRICATION

RATING: EASY

For smooth operation, the gasket must be lubricated with pure silicone lubricant. Vaseline-like in consistency, this lubricant cannot be substituted. Most other lubricants are petroleum-based which will dissolve the gasket material and cause leaks. Lubrication should be done every 6 months or when operation feels stiff. This preventive maintenance is particularly important with motorized valves since the motor will continue to fight against the sticky valve until either the diverter and shaft break apart or, more often, an expensive valve motor burns out.

Lubrication is the most important maintenance item with any three-port valve. When the valve becomes stiff to turn, it places stress on the shaft. On older models, the shaft is a separate piece that screws onto the diverter. Particularly on these models, but actually on any model, the stress from forcing a sticky valve will separate the stem from the diverter. You will be turning the handle and stem, but not affecting the

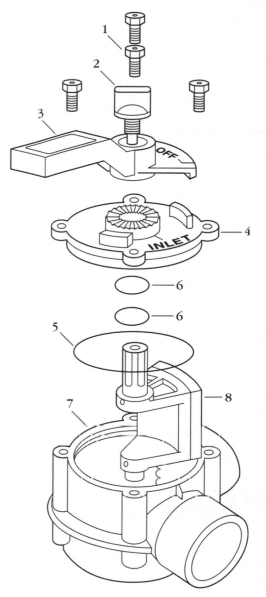

1 Cap retainer screw	5 Cap O-ring
2 Handle bolt	6 Shaft O-rings
3 Diverter handle	7 Valve body
4 Cap	8 Diverter

FIGURE 2-12 Typical three-port valve.

diverter. To repair this, remove the handle and cover, then pull out the diverter and replace it with a one-piece unit. If the gasket looks worn, replace that and lubricate it generously before reassembly.

REPAIRS

RATING: EASY

Few things go wrong with these valves, but the breakdowns that do occur are annoying and recurrent. Before disassembling any valve, check its location in relation to the spa water level. If it is below the water level, opening the valve will flood the area. You must first shut off both the suction and return lines. When installations are made below water level, they are usually equipped with shut-off valves to isolate the equipment for just such repair or maintenance work. If the valves are above the water level, you will need to reprime the system after making repairs (see the section "Priming the Pump" in Chap. 3).

Leaks are the most common failure of these valves. The valve will sometimes leak from under the cover. Either the cover gasket is too compressed and needs replacement, or the cover is loose. The cover is attached to the valve housings with metal screws. If tightened too much, the screw will strip out the hole and not be able to be tightened. The only remedy is to use a slightly larger or longer sheet metal screw to get a new grip on the plastic material of the housing. Be sure to use stainless steel screws, or else the screw will rust and break down, causing a new leak.

If this has already been done and there is not enough material left in the housing for the screw to grip, you must replace the housing. I have managed to fill the enlarged hole with super-type glue or PVC glue and, after it dries, replace the screw. These repairs usually leak and are only temporary measures. You can also fill the hole with fiberglass resin, which usually lasts longer.

Leaks also occur where the shaft comes through the cover. Remove the handle and cover, and replace the two small O-rings. Apply some silicone lubricant to the shaft before reassembly. This will lubricate the operation of the valve, decreasing friction that can wear out the O-rings. The lubricant also acts as a further sealant.

Leaks can occur where the pipes join to the housing ports. In this case, the only solution is to replace the housing. I have tried to reglue the leaking area by removing the diverter and gluing from both inside and outside of the joint. This has never worked! Try if you will, but I think you'll be wasting your time and patience.

Motorized/Automated Three-Port Valve Systems

The three-port valves just described are manually operated. These same valves can have small motors mounted in place of the manual handle for automatic or remote operation (Fig. 2-13).

The value of motorization is that the pool/spa equipment is usually located away from the pool and spa, making manual operation inconvenient. Some builders place the manual valves near the spa rather than at the equipment area; however, a small switch that operates the valve motors is often preferred. A variation of that concept is to locate the motor switch with the equipment and operate it with a remote control unit. The remote might also operate switches for lights, spa booster motors and blowers, or other optional appliances.

Installation

RATING: ADVANCED

The motorized valve in Fig. 2-13 is provided as a single unit and therefore plumbed into the system as any other valve. Other models of manual multiport valves can be easily motorized by removing the handle and the screws holding the cover in place. A mounting bracket is set on top of the cover, and slightly longer sheet metal screws (to allow for the added thickness of the bracket metal) are used to refasten the bracket and cover to the housing. These longer screws are provided with the motor bracket kits. The machine screws that normally hold the handle in place are left on the shaft. The motor unit mounts on the bracket, held in place by two screws, and the motor shaft slips over the diverter's

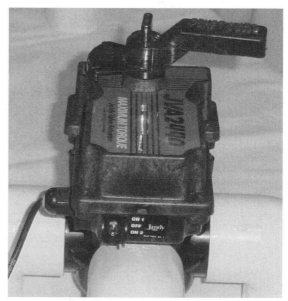

FIGURE 2-13 Motorized three-port valve.

shaft. The two shafts are secured together by tightening the screws of the diverter shaft.

Wiring diagrams are provided with each type of valve motor and are designed to operate on the standard 110 volts, 220 volts, or from an automated system that has been transformed to 12 or 24 volts. Remote and automated systems are described in greater detail in Chap. 6.

TRICKS OF THE TRADE: MOTORIZED VALVE REPAIR

RATING: ADVANCED

Few problems occur with motorized valves (beyond those discussed in the section "Manual Three-Port Valves"). As mentioned, if the valves are not properly lubricated or become jammed with debris, the motor will continue to try to rotate the valve, finally burning out. The small motors in these units are inexpensive, so if one fails, it is better to replace the unit than to attempt a repair. That said, there are a few other unique repair tips for motorized valves that are worth noting:

- To determine whether the motor has burned out, using your electrical tester, verify that current is getting to the motor. Obviously, if there is no current, the problem is in the switch or power supply and probably not the motor. If you are not familiar with basic electricity, call an electrician. If current is present, remove the motor unit from the valve and try to operate the system. If the motor rotates its shaft normally, then the problem is a stuck valve and not a burned motor. Tear down and repair the valve as described in the previous section.

- If a valve motor is burned out, it can easily be replaced without replacing the entire unit or valve. Although slightly different with each manufacturer, the process usually involves no more than four screws and three wires and will be obvious when you open the motor housing.

- Another potential problem with motorized valves is that if the mounting bracket or screws holding the unit together become loose, the unit will not align correctly with the valve. The motor will then rotate, but the valve diverter will not rotate to match. Obviously, the solution is to tighten all hardware and replace any rusted screws.

- A less frequent problem can be caused by electrolysis or simply a leaking valve. Some makes of valve motor use shafts that are made of galvanized metal or aluminum. If the valve is leaking or the motor housing is not watertight and a combination of moisture and electricity is present as a result, then the action of electrolysis will disintegrate the soft metal of the motor shaft. The motor may continue to operate, but as the shaft dissolves, it will not turn (or not completely turn) the valve diverter itself.

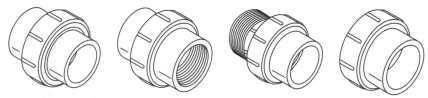

FIGURE 2-14   Typical plumbing unions.

Unions

When you need to repair or replace a pump, filter, or other equipment that is plumbed into the spa system, you must usually cut out the plumbing and replumb upon reinstallation. The concept of the union is that when you remove a particular piece of equipment, you need only unscrew the plumbing and reinstall it later in the same, simple way.

Although unions add a few dollars to your initial installation, they allow you to easily remove and replace equipment without doing any new plumbing. Unions, like other plumbing, are made of PVC plastic in standard diameters and are adapted to plumbing as any other component (glue or threaded).

Figure 2-14 shows typical plumbing unions. A nut is placed over the end of one pipe; then male and female fittings (called *shoulders*) are plumbed onto each end of the pipes to be joined. As can be seen, the joint is made by screwing the nut down on the male fitting. Teflon tape or other sealants are not needed as the design of the union prevents leaking (either by the lip design, as shown, or by an O-ring seated between the shoulders).

Unions are made for direct adaptation to spa equipment, where the pipe with the nut and female shoulder is male-threaded at its other end for direct attachment to the pump, filter, or any other female-threaded equipment. Then, only the male shoulder need be added to the next pipe, and the piece of equipment can be simply screwed into place.

Gate and Ball Valves

Gate and ball valves are designed to shut off the flow of water in a pipe and are used to isolate equipment or regulate water flow, especially where the equipment is installed below the water level of the spa.

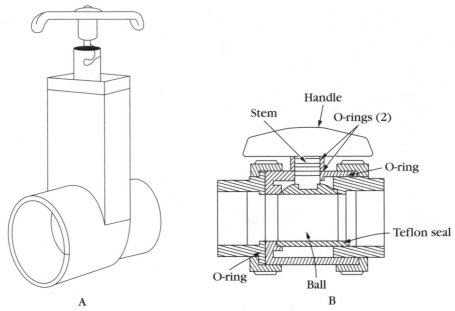

FIGURE 2-15 Shut-off valves: (A) slide style; (B) ball style.

There are basically two designs of shut-off valves. In the gate valve, as the name implies, a disk-shaped gate inside a housing slides or screws into place across the diameter of the pipe, shutting off water flow. A variation of this is the slide valve (Fig. 2-15A), where a simple guillotine-like plate is pushed into place across the diameter of the pipe.

The other design is the ball valve (Fig. 2-15B), where the valve housing contains a ball with a hole in it of similar diameter as the pipe. A handle on the valve turns the ball so the hole aligns with the pipe, allowing water flow, or aligns across the pipe, blocking flow. In all these designs, flow can be controlled by degree as well as by total on or total off.

Slide and ball valves are simple to operate and rarely break down. Repairs or parts replacement will be obvious. The gate valve, however, is operated by a handle that drives a worm-screw style shaft inside a threaded gate. If the gate sticks from obstruction and too much force is applied to the handle, the screw threads will strip out, making the valve useless. The valve cap (also called the *bonnet*) can be unscrewed, and the drive gear/gate can be removed and repaired or replaced without removing the entire valve housing. Most plumbing supply houses sell these replacement "guts," but the parts from one manufacturer are not interchangeable with those of another.

Sometimes just tightening the bonnet nut will stop leaks, but you may need to replace the O-rings on the shaft, which are designed to prevent leaks in this location.

Check Valves

The purpose of the check valve is to "check" the water flow—to allow it to go only in one direction. The uses are many and will be noted in each equipment chapter where they are employed. However, three common uses are as follows:

- In heater plumbing, to keep hot water from flowing back into the filter

- In spa air blower plumbing, to make sure air is blown into the pipe but water cannot flow back up the pipe and into the blower machinery

- In front of the pump when it is located above the spa water level, to keep water from flowing back from the filter and into the spa when the pump is shut off

There are two types of check valves. One is a flapper gate (also called a *swing valve*), and the other is a spring-loaded gate (also called a *valve seat*). The flapper style (Fig. 2-16) simply opens or closes with water flow. A diaphragm version allows water to pass on either side of a flexible membrane (Fig. 2-17).

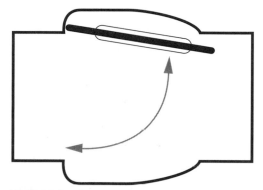

FIGURE 2-16 Flapper-style check valve.

The spring-loaded style of check valve (Fig. 2-18) can be designed to respond to certain water pressure. Depending on the strength of the spring, it may require 1, 2, or more pounds (½ to 1 millibar) of pressure before the spring-loaded gate opens. As with other valves, check valves are made of

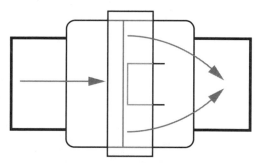

FIGURE 2-17 Diaphragm-style check valve.

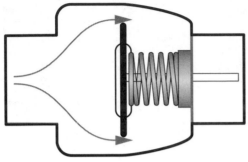

FIGURE 2-18 Spring-loaded check valve.

plastic in standard plumbing sizes and are plumbed in place with standard glue or threads.

The flapper-style valve is simple, and little can go wrong with it. It must be installed with the hinge of the flapper on top (as in Fig. 2-16). If it were installed on the bottom, gravity would leave the flapper open all the time. The only real weakness of all check valves is that they clog easily with debris, remaining permanently open or permanently closed. Because of the extra parts inside a spring-loaded check valve, it is more prone to failure from any debris allowed in the line. If the valve is threaded or installed with unions, it is easy to remove it, clear the obstruction, and reinstall.

Another solution is to use the 90-degree spring-loaded check valve, as shown in Fig. 2-19. This one allows you to unscrew the cap, remove the spring and gate, remove any obstruction, and reassemble. Be careful not to overtighten the caps—they crack easily on older models; newer models are made with beefier caps to prevent this problem. These units have an O-ring in the cap to prevent leaks. It is wise to clean these out every few months (or more frequently, depending on how dirty the pool or spa normally gets) and lubricate the gate (using silicone lube only).

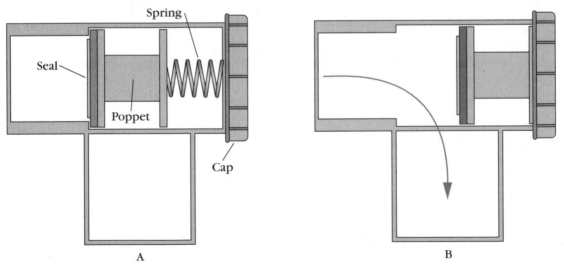

A   B

FIGURE 2-19 90-degree spring-loaded check valve: (A) closed; (B) open.

Pumps and Motors

The heart of a spa is the pump (Fig. 3-1), which is driven by an electric motor and circulates the water in the same way as the human heart speeds blood throughout the body. The proper sizing and maintenance of your spa's "heart" are essential to the safe and healthy enjoyment of your spa or hot tub.

Overview

Spa pumps are typically *centrifugal* models. That is, they accomplish their task of moving water thanks to the principle of centrifugal force (Fig. 3-2). The impeller in the pump spins, shooting water out of it. As the water escapes, a vacuum is created which demands more water to equalize this force. Thus, water is pulled from the spa and sent on its way through the plumbing and equipment.

There are two basic types of pumps used in spas—circulating pumps and booster pumps. The only significant difference between the two is that circulating pumps include a strainer basket to filter out large debris which might otherwise clog the pump or other spa equipment. Booster pumps (Fig. 3-3) typically don't include this feature, because they take water that has already been filtered in some manner and are only intended to turbocharge the return flow for the spa's characteristic massage jet action. Jetted bathtubs also use this type of pump.

FIGURE 3-1 Typical pump/motor combination.
Jacuzzi Premium.

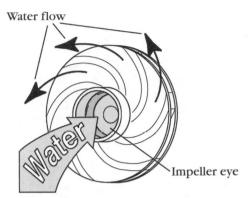

FIGURE 3-2 Centrifugal force illustrated.
Sta-Rite Industries.

FIGURE 3-3 Pump/motor without strainer pot.

Regardless of function, most spa pumps are *self-priming,* that is, they expel the air inside upon start-up, creating a vacuum that starts suction. If the pump is located below the spa's overall water level, it may not need to be self-priming, since water will be in the pump at all times anyway. Basic understanding of the components of your pump/motor unit is essential to proper maintenance and performance.

Strainer Pot and Basket

The plumbing from the main drain and/or skimmer flows to the inlet port of the pump (Fig. 3-4), usually female-threaded 1½- or 2-inch (40- or 50-millimeter) for easy plumbing. Water flows into a chamber, called the *strainer pot* or *hair and lint trap* for obvious reasons, which holds a basket (generally 4 to 6 inches or 10 to 15 centimeters in diameter and 5 to 9 inches or 13 to 23 centimeters deep) of plastic mesh that permits passage of water but traps small debris. Some baskets simply rest in the pot; others "twist-lock" in place. Most have handles to make them easier to remove. The strainer basket is similar to the skimmer basket, which traps larger debris.

On some pump models, the strainer pot is a separate component that bolts to the volute with a gasket or O-ring in between to prevent

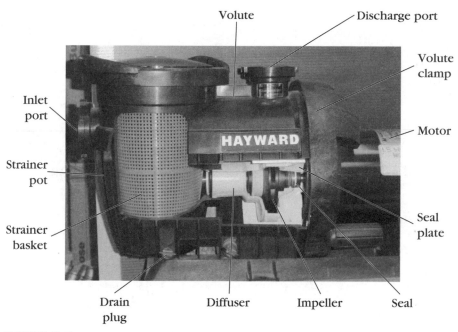

FIGURE 3-4 Cutaway view of typical spa pump.

leaks (Fig. 3-5, item 15). Sometimes the pot includes a male-threaded port which screws into a female-threaded port in the volute. In other models, the strainer pot is molded together with the volute as one piece. As noted earlier, for bathtub spas or booster pumps, where debris is not a problem, there will be no strainer pot and basket at all.

Access to the strainer basket is provided for cleaning out the debris. The strainer cover is often made of clear plastic so you can see whether the basket needs emptying. Various pump models use different methods of securing the cover, including a threaded version, a clamp, or T handles (Fig. 3-5), which screw the cover in place.

In all styles of strainer pot, the strainer cover has an O-ring that seats between it and the lip of the pot, preventing suction leaks. If this O-ring fails, the pump will suck air through this leaking area, instead of pulling water from the spa.

Notice that in the pumps of both Figs. 3-4 and 3-5 the pot has a small threaded plug which screws into the bottom. This plug is designed to allow complete drainage of the pot for winterizing the pump. It also is made of a weaker material than that of the pot (on metal pots, for

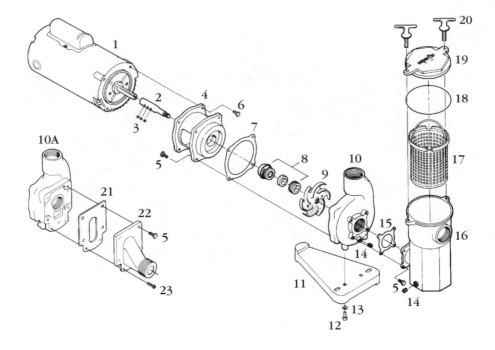

1 Motor (C frame)
2 Shaft extender
3 Allen setscrew
4 Bracket
5/6 Hex bolt
7 Volute gasket
8 Shaft seal
9 Impeller
10 Volute
10A Alternate volute
(for use without strainer pot)
11 Base

12 Hex bolt
13 Lock washer
14 Drain plug
15 Trap gasket
16 Strainer pot
17 Strainer basket
18 Lid O-ring
19 Strainer pot lid
20 T-handle
21 Gasket
22 Suction flange
(for use without strainer pot)
23 Assembly screw

FIGURE 3-5 Exploded view of typical bronze pump and motor. *Aqua-Flo, Inc.*

example, the plug is made of plastic, soft lead, or brass). If the water in the pot should freeze, this "sacrificial" plug will pop out as the freezing water expands, relieving the pressure in the pot. Otherwise, of course, the pot itself would crack.

Volute

The volute is the chamber in which the impeller spins, forcing the water out of the pump and into the filter (or directly back to the spa if the system is not filtered or heated). The outlet port is usually female-threaded 1½ or 2 inches (40 or 50 millimeters) for easy plumbing. When the impeller moves water, it sucks the water from the strainer pot.

The impeller by itself will move the water, but it cannot create a strong vacuum to start the water flow. The area immediately around the impeller must be limited to eliminate air and help start that flow. A diffuser and/or closed-face impeller helps this process, but on many pump designs, the volute serves this purpose.

Figure 3-4 shows a pump design with a separate diffuser which houses the impeller. In some designs, the inside of the volute is ribbed to improve the flow efficiency. On other models, the volute itself closely encircles the impeller to create this diffuser effect without a separate component (Fig. 3-5).

Impeller

The impeller is the ribbed disk (these curved ribs are called *vanes,* and the disk is called a *shroud*) that spins inside the volute. Take another look at Fig. 3-2—as water enters the center or "eye" of the impeller, it is forced by the vanes to the outside edge of the disk. As the water is moved to the edge, there is a resulting drop in pressure at the eye, creating a vacuum which is the suction of the pump.

The amount of suction is determined by the design of the impeller and pump components and the strength of the motor that spins the impeller.

Most impellers on spa pumps have a female-threaded hole on the center backside (the side facing away from the water source) and screw onto the male-threaded end of the motor shaft. The rotation of the shaft is just like the action of a bolt being threaded into a nut. Therefore, as the shaft turns, it is always tightening the impeller on itself.

Impellers are rated by horsepower to match the motor horsepower that is used. These, in turn, determine the horsepower rating of the pump or pump/motor you have.

Seal Plate and Adapter Bracket

To remove the impeller or to access the seal, the volute is actually divided into two sections—the actual curved housing for the impeller is actually the volute, while its other half is called the *seal plate* or *adapter bracket.*

The seal plate is joined to the volute with a clamp. An O-ring or gasket between them makes this joint watertight. The motor is bolted directly onto this type of seal plate. In other designs, the seal plate is molded together with an adapter bracket that supports the motor and bolts to the volute with a paper or rubber gasket between them to create a watertight joint.

Some pumps use a clamp that is made tight by a bolt and nut. Always assemble the clamp with the bolt on top to make future access easier. Moreover, if the pump leaks at all, the bolt gets wet, causing it to rust, making unscrewing it even tougher, or the bolt breaks altogether.

In all pump types, the shaft of the motor passes through a hole in the center of the seal plate, and the impeller is attached, threaded onto the shaft.

Seal

If the shaft passed through the large hole of the seal plate without some kind of seal, the pump would leak water profusely. If the hole were made small and tight, the high speed of the shaft spinning would create friction heat and burn up the components in minutes, or the shaft would bind and not turn at all. Some clever engineer devised a solution to this problem, called a *seal.*

The seal allows the shaft to turn freely while keeping the water from leaking out of the pump. The seal has two parts. In Fig. 3-4, the right half of the seal is composed of a rubber gasket or O-ring around a ceramic ring. This assembly fits into a groove in the back of the impeller. The ceramic ring can withstand the heat created by friction. The left half of the seal fits into a groove in the seal plate and is composed of a metal bushing containing a spring. A heat-resistant graphite facing material is added to the end of the spring which will face the ceramic in the other half.

The tight fit of the seal halves prevents water from leaking out of the pump. The spring puts pressure against the two halves to prevent them from leaking. As the shaft turns, these two halves spin against each other but do not burn up because their materials are heat-resistant and the entire seal is cooled by the water around it. Therefore, if the pump is allowed to run dry, the seal will be the first component to overheat and fail.

Pump makers are always improving the heat dissipation (called heat *sink*) capabilities of their pumps, so that dry operation will result in little or no damage to the seal or pump components. Still, no matter what the maker claims, pumps are not designed to run without water for more than a few minutes while priming.

Motor

Motors, like the pumps they drive, are rated by horsepower, typically ½, ¾, 1.0, 1.5, and 2.0 horsepower. The caps on each end of the motor housing are called *end bells.* A starting switch is mounted on one end with a small removable panel for connection and maintenance access.

In here you will also find the *thermal overload protector.* This heat-sensitive switch is like a small circuit breaker. If the internal temperature gets too hot, it shuts off the flow of electricity to the motor to prevent greater damage. As this protector cools, it will automatically restart; but if the unit overheats again, it will continue to cycle on and off until the problem is solved or the protector burns out altogether.

It takes a great deal of electricity to start a motor but far less to keep it going (in fact, about 5 to 6 times as much). The capacitor, as the name implies, has a capacity to store an electric charge. The capacitor is discharged to give the motor enough of a jolt to start; then the motor is able to run on the lower operating amount of electricity as designed. Without the capacitor, the motor would need to be served by very heavy wiring and high-amp circuit breakers to carry the starting amps. Capacitors are located in a separate little housing mounted atop the motor housing or inside the front end bell.

Some motors are designed to operate at two speeds. For example, some spas operate at high speed for jet action, but lower speeds for circulation and heating. Normal rotation speed is 3450 revolutions per minute (rpm), and the low speed is 1750 revolutions per minute.

Most motors are designed to be connected to either a 110-volt or a 220-volt power source. By changing a wire or two internally, you determine which voltage is used. The instructions for this conversion are printed on the motor housing or on the inside of the access cover.

Finally, the housing of the motor is designed to adapt to various kinds of pump designs, called *square flange, C-frame,* the *48,* the *uniseal flange,* and others. When you replace a motor, be sure to buy the design that fits your pump.

Choosing the Right Pump and Motor

In most cases we can just assume the original designer or builder used the correct size pump and motor for the job and make our replacement accordingly. But what if the original equipment was too small or too large? What if the plumbing has been repaired (which may have added or deleted pipe and fittings) or equipment has been added or deleted, thus changing the system and requiring the pump to work more (or less)? What if the identifying rating plates have been removed or are so weather-worn that now we cannot tell what size the existing equipment might be? Finally, what if this is a brand new installation? How can we choose the right pump/motor for the job?

Well, I'm glad you asked all those intelligent questions. The answer is that we need to know a little about the spa system's needs and hydraulics. A detailed explanation of pump and plumbing hydraulics can be found in another of my books, *The Ultimate Pool Maintenance Manual* (2d ed., McGraw-Hill, 2001). Most spa owners will not need the information in such detail, but it is important to understand the turnover rate of your spa.

Turnover Rate

The *turnover rate* of the body of water reflects how long it takes to run all the water through the system. It is desirable for the water to completely circulate through the filter at least one to two times per day, but local codes will generally require a specific time period. In Los Angeles, for example, commercial spas must turn over in half an hour, and pumps must be run for several hours each day to ensure clean water. That's a pretty good rule of thumb for a spa in your home, too, since the goal of any turnover rate is to keep the water clean and healthy.

Your pump will be labeled with a rating in gallons per minute (gpm) or liters per minute (lpm). This number will vary if you change the hydraulics significantly, by relocating the pump far away from the spa or by adding jets on the discharge side of the plumbing. Assuming that you haven't changed the hydraulics, you can consult the owner's manual for the pump and see the average rating for a typical spa. A little math will tell you if your pump, or the new one you intend to install, is adequate for the intended turnover rate.

Let's say the spa holds 1000 gallons (3785 liters) of water. The calculations look like this:

1000 gallons ÷ 30 (remember—half an hour
to turn over all the water) = 33 gpm (125 lpm)

In other words, if you have a pump that can deliver at least 33 gpm (125 lpm), then you will have an adequate turnover of your spa water. Of course, as filters become clogged and dirty or debris in the skimmer or strainer baskets slows the flow, you may want a more powerful pump to keep up with the 30-minute turnover rate goal. You might look for a pump that can deliver up to 50% more gallons per minute (in this example, around 50 gpm or 189 lpm) to ensure that your circulation remains adequate under all conditions.

Another consideration in choosing the right pump for your spa is the number of jets. As a general rule of thumb, to operate efficiently, spa jets require 15 gpm (57 lpm) running through each one. Therefore, if you have a system that delivers 60 gpm (227 lpm), you can install up to four jets. As another generality, each jet requires ¼ horsepower from its pump/motor, so that four-jet spa would need at least a 1-horse-power unit.

Remember, this assumes the pump is doing no other work. If it is pushing water through the filter and heater before it gets back to the jets, or if the equipment is more than 20 feet (6 meters) from the spa, then some power will be lost, so you will need to calculate more than ¼ horsepower per jet. If you have any doubts or find that the pump isn't moving the water as expected, consult a pool or spa technician to calculate the hydraulic pressures in your system and determine the correct pump for you.

Booster Pumps

All the information provided in this chapter applies equally to circulating pumps or booster pumps, but booster pumps do have some unique properties. As already mentioned, some spas use the circulating pump to also power the jets, but other installations have a second pump that is not restricted by the other components.

The booster pump should be plumbed with its own suction drain and plumbing, so that it is not competing with the circulation pump, one starving the other for water flow. Some main drains are designed with two ports under one cover, so you can connect a suction line for each pump. This may be preferable to adding a second drain fitting in a small spa.

Similarly, don't try to force the return lines together into a few jets. The system can handle only so much water, so plumb some jets to one pump and others separately to the other. Finally, if a booster pump is likely to pull in debris, it will require a strainer basket just as the circulation pump does. In spas, hair is the most common occupant of either pump strainer basket, but since the circulation pump is connected to the skimmer, it is more likely to contain leaves or other debris, while the booster pump strainer basket collects mostly hair and grit.

Portable spas typically use two-speed motors on a single pump, instead of two separate pumps. The low speed is used for circulation and heating, the high speed for jet action.

Maintenance and Repairs

Since the pump and motor are the heart of the system, pumping water through the other components, if they don't perform efficiently, the abilities of the other components won't much matter.

Keeping the motor in good working order is a matter of keeping it dry and cool. The best detection tool of motor problems will be your ears. Laboring motors or those with bad bearings will let you know quickly (see "Troubleshooting Pumps and Motors" later in this chapter).

Keeping the pump in good order is more a matter of sight. Leaks will alert you that the pump needs attention. Here are the basic repairs and maintenance of the pump/motor unit, starting from the front, the first place the water encounters.

TOOLS OF THE TRADE: PUMPS AND MOTORS

- Flat-blade screwdriver
- Phillips screwdriver
- Allen-head wrench set
- Open end/box wrench set
- Hacksaw
- $\frac{5}{16}$-inch (8-millimeter) nut driver
- Impeller wrenches
- Teflon tape
- Silicone lube
- Needle-nose pliers
- Hammer
- Seal driver
- Emery cloth or fine sandpaper
- Impeller gauge

Strainer Pots

RATING: EASY

Clean out the strainer basket often. Even small amounts of hair or debris can clog up the fine mesh of the basket and can reduce flow substantially. This simple preventive maintenance and keeping a clean skimmer basket are the two simplest and most important elements in keeping a clean spa and all the components working. If the water can't flow adequately, it can't filter or heat adequately either.

1. **Prepare** Shut off the pump and close the valves (if the pump is below the water level) to avoid flooding.

2. **Clean** Remove the strainer pot cover and clean out the basket.

3. **Prime** If the pump is above the water level of the spa, fill the pot so the pump will reprime easily.

4. **Close** Check the cover O-ring for nicks or breaks before replacing the cover. Tighten the bolts or clamps, open the valves, and restart the system.

5. **Look** Check that water is circulating freely after the air has been purged from the system.

The only other problems you might encounter at the strainer pot are broken baskets or a crack in the pot itself. If the basket is cracked, it will soon break, so replace it. If allowed to operate with a hole in it, the basket will permit large debris and hair to clog the impeller or some of the plumbing between the equipment components.

Cracks might develop in the pot itself, especially if you live where it gets cold enough to freeze the water in the pot. Again, the only remedy is replacement. Remove the entire pump and motor assembly and take it to your spa professional to replace the pot. More detailed repair procedures can also be found for the do-it-yourselfer in *The Ultimate Pool Maintenance Manual.*

Gaskets and O-Rings

Most problems occur in strainer pots when the pump is operated dry. The air heats up as the impeller turns without water to cool it. The strainer basket will melt; the pot cover, if plastic, will warp; and the O-ring will melt or deform. Usually, replacement of the overheated parts will solve the problem.

GASKETS

RATING: ADVANCED

When gaskets leak, or in extreme cases, if the strainer pot itself must be replaced, the procedure is quite simple:

1. **Disassemble** Remove the strainer pot. Take out the four bolts (Fig. 3-5, item 5), usually using a ½- or ⁹⁄₁₆-inch box wrench.

2. **Clean** Clean off the old gasket thoroughly. Failure to do this will leave gaps in the new one that will eventually leak.

3. **Reassemble** Reassemble the pump with a new gasket (Fig. 3-5, item 7) in the same way the old one came off. Tighten the bolts evenly

(so the new gasket compresses evenly) by gently securing one bolt, then the one opposite, and then the last two. Continue tightening in this criss-cross pattern until each bolt is hand tight. When you are dealing with plastic pumps, overtightening will cause the bolt to crack the pump components or strip out the female side.

Some pumps are designed with bolts that go through the opening in the pot and volute and are secured by a nut and lockwasher on the other side. Still, do not overtighten, as you will crack the pump components. The key to this simple procedure, as with virtually all other mechanical repair, is to carefully observe how the item comes apart. It will go back together the same way!

O-RINGS

RATING: EASY

When you are removing or replacing the strainer pot cover, be sure the O-ring and top of the strainer pot (Fig. 3-5, items 18 and 19) are clean, as debris can cause gaps in the seal. Sometimes these O-rings become too compressed or dried and brittle, and they cannot seal the cover to the pot. In this case, replace the O-ring.

If no replacement is available, try turning the O-ring over. Sometimes the rubber is more flexible on the side facing the cover. Be careful to remove the O-ring gently. Too much stress will cause the rubber to stretch out, making it too large around to return to the groove in the cover. If it has stretched out, try soaking it in ice water for a few minutes to shrink it. Finally, coat the O-ring liberally with silicone lube. This can take up some slack and complete the seal if the O-ring is not too worn out.

Another emergency "trick" is to run Teflon plumbing tape around the O-ring to give it greater bulk and make it seal. If you use this trick, be sure to wind the tape evenly and tightly around the O-ring, so that loose or excess tape does not cause an

---

TRICKS OF THE TRADE: PUMP AND MOTOR HEALTH CHECKLIST

RATING: EASY

Look

- Motor dry
- Vents free of leaves or other debris
- No pump leaks
- Strainer pot clean

Listen

- Steady, normal hum
- No laboring, cavitating, or grinding noises

Feel

- Motor warm, but not hot
- No major vibration

even worse seal. If the O-ring has actually broken, it will almost always leak at that spot. However, I have used the Teflon tape trick successfully in these cases for a temporary repair when a new O-ring was not immediately available.

Pump and/or Motor Removal and Reinstallation

RATING: ADVANCED

Sometimes it is necessary to remove an entire pump/motor unit to take it apart or complete a repair. If the pump is damaged beyond your ability to repair it, you may want to take the entire unit to a motor repair shop. The shop can rebuild it as needed, then you reinstall it on the job site. Your local pool/spa supply house can recommend a rebuilder, or you can consult the phonebook.

1. **Cut the Plumbing** Disconnect the plumbing. Most spa pump/motor combinations will be installed with threaded unions, making removal quite simple. Other spas require that you cut the plumbing on the suction and return side of the pump to remove the unit. In these cases, cut the pipe (Fig. 3-6A and B) with enough remaining on each side of the cut to replumb it later. Ideally, a few inches (or millimeters) on each side will allow you to use a slip coupling to simply reglue the unit in place later (as described in Chap. 2).

2. **Disconnect the Electrical Wiring** Some spas even provide a plug-in connector for the electrical supply, making this part of the job easy, too. But in some models, you may have to disconnect a hardwired electrical connection for your pump/motor:

    A. Shut off the electrical circuit breaker to be sure no current can get to the spa equipment while you are working.

    B. Remove the access cover to the switch plate area of the motor (Fig. 3-6C), near the hole where the conduit enters the motor.

    C. Remove the three wires inside the motor and unscrew the conduit connector (Fig. 3-6D) from the motor housing. You can now pull the conduit and wiring away from the motor, and the entire pump/motor should be free.

    D. There may be an additional bonding wire (an insulated or bare copper wire that bonds or grounds all the equipment together

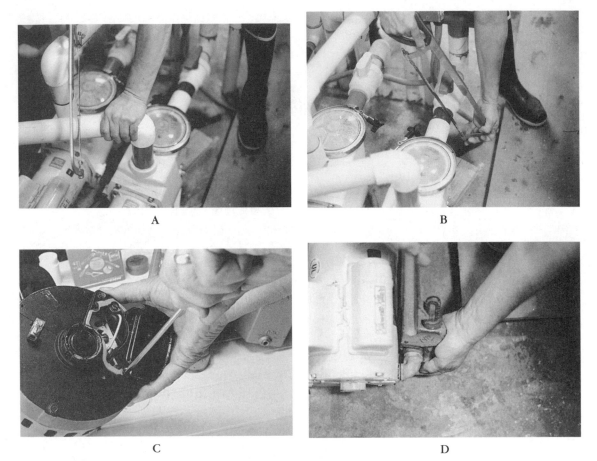

A   B

C   D

FIGURE 3-6 Removing pump/motor unit: (A, B) plumbing; (C, D) electrical.

and to a grounding system). This is easily removed by loosening the screw or clamp that holds it in place.

E. Tape off the ends of the wires, even though the breaker is shut off, and put tape over the breaker switch itself. Leave a note on the breaker box to yourself, family members, or the customer to be sure no one accidentally turns the breaker back on while the pump/motor is away for service.

3. Unbolt the Base Pumps should be secured to the spa cabinet floor or the concrete base of your equipment area. Simply unscrewing the bolts will release the unit.

Simply reverse the order of these steps when you return the pump to service. When you reinstall the plumbing, take a look at the equipment area. Keep bends and turns to a minimum. Remember, each turn creates resistance in the system. Also, don't locate the pump close to the base of the filter. When you open the filter for cleaning, water is sure to flood the motor. Last, try to keep motors at least 6 inches (15 centimeters) off the ground. The bracket of the pump does this in part, but heavy rains or flooding from broken pipes and filter cleanings can flood the motor if it is too close to the ground.

Use this basic removal and reinstallation procedure whenever there is a problem you can't solve, such as replacing seals or motor components. It will be faster and cheaper in most cases to bring the unit to your spa professional for more complex repairs, but you will save a lot of money by avoiding the house call by a technician for the removal and reinstallation.

Troubleshooting Pumps and Motors

RATING: EASY

The first and most common motor ailment is water! Motors get soaked in heavy rain, when you take the lid off the filter for cleaning, when a pipe breaks, or when you look at it wrong! In all cases, dry it off and give it 24 hours to air-dry inside as well before you start it up—moisture on the windings will short them out and short out your warranty as well. The basic problems beyond this are as follows.

MOTOR WON'T START

Check the electrical supply at the time clock and breaker panel, and look for any loose connection of the wires to the motor. Sometimes one of the electrical supply wires connected to the motor switch plate has become dirty. Dirt creates resistance which creates heat which ultimately melts the wire, breaking the connection. Clean the switch plate terminals, too, and reconnect the wiring.

MOTOR HUMS BUT WON'T RUN

Either the capacitor is bad or the impeller is jammed. Open the strainer pot and remove the basket. Use one finger to manually turn the impeller. If it won't spin, there is an obstruction and the pump must be disassembled to remove it. If it does spin, the capacitor is probably bad and you should call your spa professional.

LOUD NOISE OR VIBRATIONS

First check that the pump/motor unit is firmly bolted to the floor or cabinet. If it is secure, the problem is likely to be worn out bearings. Remove the unit and have the motor replaced.

THE BREAKER TRIPS

Use the procedure described above to disconnect the motor's electrical supply and then reset the breaker. Turn the motor switch (or time clock "on" switch) back on; if it trips again, the problem is either a bad breaker or more likely bad wiring between the breaker and the motor. Be very careful with this test, of course. Switching the power back on with no appliance connected means you are now dealing with bare, "live" wires. Be sure no one is touching them and that they are not touching the water, each other, or anything else. If the breaker does not trip when you conduct this experiment, the motor is bad and will need to be replaced.

Priming the Pump

RATING: EASY

Sometimes the most difficult step is to get water moving through the pump. *Priming* means getting water started, creating a vacuum so more water will follow. If the pump is located below the water level of the spa itself, water will always flow normally into the pump. If the pump is located above that water level or some distance from the spa, you may need to assist the priming process:

1. **Check the Water Level** Before starting a pump that you have had apart, always make sure there is enough water in the spa to supply the pump. When you take apart equipment, water is usually lost in the process, and now there may not be enough to fill the skimmer.

2. **Check the Water's Path** Often, priming problems are not related to the pump, but to some obstruction. Check the main drain and the skimmer throat for leaves, debris, or other obstruction. Open the strainer pot lid, remove the basket, and make sure there are no obstructions or clogs in the impeller. Last, make sure that once the

pump is primed, it has somewhere to deliver the water! In other words, be sure all valves are open and that there are no other restrictions in the plumbing or equipment after the pump.

3. **Fill the Pump** Always fill the strainer pot with water and replace the lid tightly so air cannot leak in. Keep adding water until the pot overflows, so you fill as much of the pipe as the pot.

TRICKS OF THE TRADE: NOISE CHECKLIST

RATING: EASY

Not all noise is caused by the motor. Track down noises by a process of elimination, experimenting with various pieces of equipment (such as booster motors, automated valves, spa blowers, and heaters) all turned off, then turned on one at a time. Here's a list that will help you find other culprits:

Security

- Is the pump properly secured to the deck or mounting block, and is the mounting block secure?
- Are check valves rattling?
- Are pipes loose and vibrating? Hold onto sections of exposed pipe and see if the noise changes.

Cavitation

- Are suction and return line valves fully open or open too much?
- Is suction plumbing undersized? Refer to the hydraulics section.

Air

- Is the pump strainer basket clean and the lid tightly fastened?
- Is the skimmer clogged or the water level low?

Other troublemakers

- Is the equipment located in a sound-magnifying environment, such as large concrete pad and masonry walls? Consider a vented "doghouse" cover.
- Is the heater "whining"? (See Chap. 5, "Heaters.")
- Is the spa air blower loose or vibrating, or is the discharge restricted, producing a louder sound?
- Are loose filter grids rattling inside the filter canister?

4. **Start the Motor** Turn the pump motor, and open the air relief valve on top of the filter (see Chap. 4 to locate this valve). Give the pump up to 2 minutes to "catch" prime.

5. **If at First You Don't Succeed…** If water does not begin flowing, repeat the steps. If that fails, you may have an air leak. The pump is sucking air from somewhere, meaning it will not suck water. Air leaks are usually in strainer pot lid O-rings, or the pot or lid itself has small cracks. The gasket between the pot and the volute may be dried out and leaking. Of course, plumbing leading into the pump might be cracked and leaking air. If any of these components leak air in, they will also leak water out. When the area around the pump is dry, carefully fill the strainer pot with water and look for leaks out of the pot, volute, fittings, or pipes. Another method is to fill and close the pot, then listen for the "sizzling" sound of air being sucked in through a crack as the water drains back to the spa.

Cost of Operation

As with any electrical appliance, you can easily calculate the cost of operation. Electricity is sold by the kilowatt-hour. This is 1000 watts of energy per hour. We know that volts × amps = watts, so we look at the motor nameplate. We see that the motor runs, for example, at 15 amps when supplied with 110-volt service and at 7 amps when supplied by 220-volt service.

Let's say the service in our example is running on 220-volt service. We multiply 220 (volts) × 7 (amps) = 1540 watts. Looking at a local electric bill, we learn that we pay 15 cents per kilowatt-hour. As noted above, 1 kilowatt is 1000 watts, so if we divide 1540 watts by 1000, we get 1.54. That is multiplied by our kilowatt rate (of 15 cents), which equals 23 cents for every hour we run the appliance.

If we run the motor 2 hours per day, that means 23 cents × 2 hours = 46 cents per day. Over 1 month, that equals 30 × $0.46 = $13.80.

Filters

I f the pump is the heart of the spa equipment system, then the filter is the kidney, and since there are no moving parts, it is the easiest piece of equipment to maintain.

Filter Types

All filters are basically the same—a canister containing some type of filter *medium.* This media strains impurities out of the spa water as it passes through. There are essentially three types of filter media, but one is most commonly used for spas.

Cartridge Filters

As the name implies, the medium used in this type of filter is a *cartridge* made of fine-mesh, pleated fabric (usually polyester). Figure 4-1A shows a cartridge filter for a typical residential spa, and Fig. 4-1B shows the same filter built right into the spa. However, larger versions of this filter can contain one or more cylindrical cartridges to handle larger bodies of water (Fig. 4-2).

Water enters the metal or fiberglass filter canister and flows over the cartridge(s). The extremely tight mesh of this fabric strains impurities out of the water before water is forced into the center of the cartridge, finally leaving the unit through plumbing at the base.

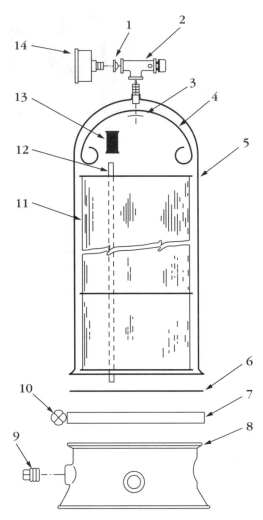

1 Screen filter
2 Air valve
   assembly
3 Retainer
4 Brace
5 Housing
6 O-ring
7 Clamp
8 Base
9 Drain plug
10 Clamp bolt
    assembly
11 Filter cartridge
12 Air bleed tube
13 Air tube screen
14 Pressure gauge

A

B

FIGURE 4-1 **(A) Typical cartridge filter. (B) In-spa cartridge filter.** *A: Sta-Rite Industries.*

Cartridge filters are classified by square footage of filter surface. By pleating the cartridge material, a great deal of square footage can fit into a very small package. Typically, cartridge filters for residential spas range from 20 to 120 square feet (2 to 12 square meters) in a tank not more than 4 feet (1.2 meters) high by 1½ feet (46 centimeters) in diameter.

Other Filter Types

You might encounter one of the other two types of filter media in a spa, especially if it is part of a pool/spa combination, a wooden hot tub, or a large commercial spa. Although cartridge filters work equally well on spas and hot tubs, some wooden tub owners prefer a filter that is less likely to clog with the fine cellulose that can strip off the wood and become lodged in the filter. In fact, very little of that stringy wood pulp

FIGURE 4-2 Cutaway view of multiple-cartridge filter.

should end up in any filter—an excess of pulp is a sign of a poorly maintained tub. Nonetheless the other two types of filter you might encounter use diatomaceous earth (DE) or sand as the filter medium.

DIATOMACEOUS EARTH FILTERS

In the diatomaceous earth type of filter, also called a DE filter (Fig. 4-3A), water passes into a metal or plastic tank, through a series of fabric-covered grids (also called *filter elements*), and back out of the unit. The grids do not actually perform the filtration process, but instead are coated with a filtration medium, diatomaceous earth, which does the actual filtering work.

DE is a white, powdery substance found in the ground in large natural deposits, actually composed of the skeletons of billions of microscopic organisms that were present on earth millions of years ago. If you look at DE under a microscope, you see what appear to be tiny sponges, so the filtration ability becomes more apparent. Just as with a sponge, water can pass through, but impurities in the water cannot. Because the DE

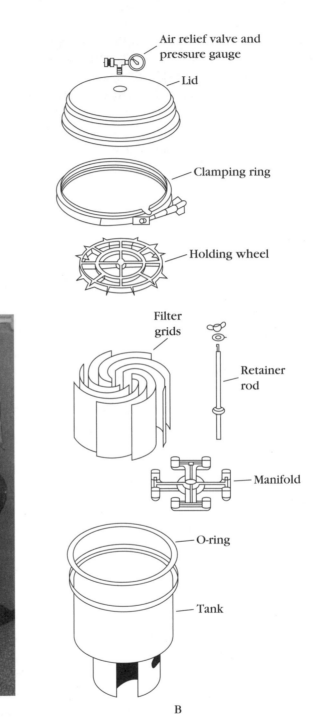

FIGURE 4-3 (A) Cutaway view of typical DE filter. (B) Exploded view of DE filter.

particles are so fine, they can strain very small particles from the water as it passes through.

As can be seen in Fig. 4-3B, the grids are mounted on a manifold, and the resulting assembly fits into the tank. A retaining rod through the center screws into the base of the tank, and a holding wheel keeps the grids firmly in place. The top of the tank is held in place by a clamping ring, the two parts sealed with a thick O-ring to prevent any leaking.

The water enters the tank at the bottom and flows up around the outside of the grid assembly. It must flow through the grids, down the stem of each grid, and into the hollow manifold, after which it is sent back out of the filter.

One partial way to clean this filter is called *backwash,* a concept that we will discuss in greater detail later. As the term suggests, backwashing means the water is redirected through the filter in the opposite direction to normal filtration (accomplished with a backwash valve, also discussed below), thereby flushing old DE and dirt out of the filter.

Not all vertical DE filters are equipped with a backwash valve to allow backwash-type cleaning and so must be disassembled each time for cleaning. Some are also equipped for "bumping" rather than backwashing. In this process, the dirty DE is bumped off the grids, mixed inside the tank so the dirt is evenly distributed within the DE material, then recoated onto the grids.

The bumping concept is this: Since most dirt sits on the outermost layer of DE (as the DE rests on grids), there is still relatively clean, unused DE available beneath. By mixing DE and dirt together, somewhat cleaner DE will be brought to the surface, preventing the need to actually replace it. Bumping should only be used as a temporary filter method, however, since it is not very effective at cleaning the filter.

In some areas, out of concern that DE will clog pipes when the grids are cleaned or backwashed, local codes require that DE not be dumped into the sewer system. In this case, a *separation tank* is added next to the filter. When you are draining or backwashing the filter, the dirty water is passed through a canvas bag inside a small tank before going into sewer or storm drains. The canvas bag strains most of the DE out of the water so it can be disposed of elsewhere.

Like cartridge filters, DE filters are sized by the square footage of surface area of their filtration media. In DE filters, therefore, the total surface square footage of the grids is the size of the filter. Typically, there are 8 grids in a filter totaling 24 to 72 square feet (2 to 7 square meters), designed into tanks of 2 to 5 feet (60 to 150 centimeters) high by around 2 feet (60 centimeters) in diameter. Obviously the larger the filter, the greater the capacity it will have to move water through it. Therefore, filters are also rated by how many gallons or liters per minute can flow through them.

SAND FILTERS

Another filtration medium is common sand, which is the natural method of filtering water. For spas, the most common sand filter is the *high-rate* sand filter, and it is quite simple to understand.

Water is passed through a layer of sand and gravel inside a tank, which strains impurities from the water before it leaves the tank. As with DE filters, the water is under pressure inside the tank from the resistance created by trying to push it through the filtration media. This differentiates it from another type of sand and gravel filter, used especially in fish ponds where there is no such pressure, the *free flow* sand filter. Since free flow sand filters are not typically used for spas or hot tubs, these will not be discussed further.

Figure 4-4A shows a spa equipment package that includes a sand filter. Water enters the tank through the valve on top and sprays over the sand inside. Water runs through the sand, with the impurities being caught by the sharp edges of the grains, and is pushed through the manifold at the bottom where it is directed up through the pipe in the center and out of the filter through another port of the valve on top. The individual fingers of the drain manifold are called *laterals* (Fig. 4-4B), and the center pipe used to return clean water is called a *stanchion pipe.* To drain the tank, a drainpipe is provided at the bottom as well.

Like other types of filters, sand filters are sized by square footage (or square meters) and the resulting ability of the unit in gallons or liters per minute. Knowing the volume of sand recommended for any given filter (expressed in cubic feet), the manufacturer will arrive at a square footage (or square meter) value and a resulting gallons or liters per minute rating. Sand filter tanks are usually large, round, fiberglass units, between 2 and 4 feet (60 and 120 centimeters) in diameter.

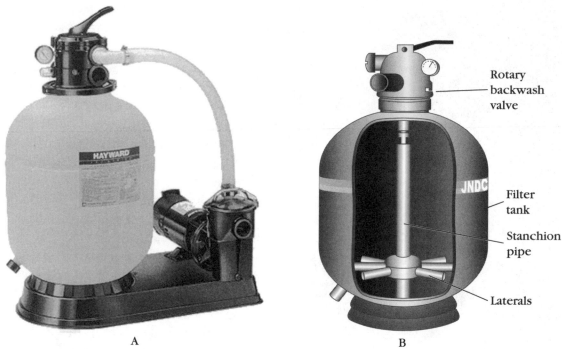

FIGURE 4-4   (A) Typical sand filter. (B) Interior view of sand filter.

Filter Components

Regardless of the filtration media, all filters will have some hardware in common. These include pressure gauges and air relief valves. Sand and DE filters also have backwash valves.

Pressure Gauges and Air Relief Valves

Most filters of all types are fitted with a pressure gauge, mounted on top of the filter, often in combination with an air relief valve (Fig. 4-5). The pressure gauge shows 0 to 60 pounds of pressure per square inch (0 to 4000 millibars) and is a useful tool in several ways.

First, when a filter is first put into service or has just been cleaned, I make it a habit to note the normal operating pressure. In fact, I carry a waterproof felt marking pen and write that pressure directly on the top of the filter. Most manufacturers tell you that when the pressure increases more than 10 pounds per square inch (700 millibars) over this normal operating pressure, it is time to clean the filter again.

FIGURE 4-5 Pressure gauge and air relief valve.

The other benefit of the pressure gauge is that it enables you to quickly spot operating problems in the equipment system. If the pressure is much lower than normal, something is obstructing the water coming into the filter—if it can't get enough water, it can't build up normal pressure. If the gauge reads unusually high, either the filter is dirty or there is some obstruction in the flow of water after the filter.

When the pressure fluctuates while the pump is operating, the spa may be low on water or have some obstruction at the skimmer—when the water flows in, the pressure builds; then as the pump sucks the skimmer dry, the pressure drops off again. This cycling will repeat, or the pressure will simply drop altogether, indicating the pump has finally lost prime. In any event, a pressure gauge is a valuable instrument on the filter.

You will normally find an air relief valve mounted on a yoke or T fitting along with the pressure gauge. This is simply a threaded plug that, when loosened, allows air to escape from the filter. Purging the air is important to proper filtration, because the air can displace water, essentially reducing the filter area of the sand, DE, or cartridge medium. When a circulation system first starts up, particularly after cleaning or if the pump lost prime, there is a lot of air present in the filter. If the air isn't purged, the filter may be operating only at one-half its capacity.

Figure 4-1 shows an inside view of a cartridge filter. The area above the cartridge is called the *freeboard.* This empty area is present above the filtration media of all filters. Air in this area is quite acceptable, since there is no filtration medium in the freeboard.

Thus, it is important to let the air out of the filter at least once each week. Again, some air is normal, so don't be obsessive about this. Just be sure that the filtration medium itself is covered with water in the tank, not air.

Backwash Valves

As noted previously, backwashing is a method of cleaning a DE or sand filter (cartridge filters do not backwash) by running water backward through the filter, flushing the dirt out to a waste drain line or sewer line. There are two types of backwash valve—the piston and the rotary (with the multiport valve being a variation of the rotary valve). Since few filters today use the piston backwash valve, we will focus our attention on the more common, rotary type.

Some filters use a rotary backwash valve like the one in Fig. 4-4A. Changing the direction of water flow is accomplished by rotating the interior rotor. A rotor gasket seal or O-rings keep water from leaking into the wrong chamber.

Other filters use top- or side-mounted backwash valves (Fig. 4-6), while DE filters typically mount the valve underneath the tank. In any of these designs, a handle allows you to rotate the interior rotor to align with the openings of the valve body as desired.

In normal filtration, water comes into the valve through the opening marked "Intake" or "From pump discharge." After passing through the filter media, the water flows on to the heater or back to the spa.

When the valve is rotated into the "backwash" position, water is sent in the opposite direction, removing dirt (and DE, if the filter uses that medium). The flow is then directed to the opening marked "Backwash," which is connected to a waste or sewer line. Many backwash discharges are not hard-plumbed, and you must connect a hose to direct

FIGURE 4-6 Backwash valve.

the wastewater wherever you want it to go. Normally a cheap, blue, collapsible plastic backwash hose is attached to the waste opening with a hose clamp. This 1½- or 2-inch-diameter (40- or 50-millimeter) hose is intentionally made rather flimsy, since the water will not be under much pressure when draining to waste and to allow the hose to be easily rolled up and stored near the filter. This ease of rolling up is an advantage since backwash hoses normally come in lengths of 20 to 200 feet (6 to 60 meters)—you may have to route wastewater into a street storm drain, so very long lengths are common.

By the way, dirty filter water is an excellent fertilizer for lawns as it is usually rich in biological nutrients, algae, decaying matter, and DE (which, as you now know, is a natural material). You may run your backwash hose on the lawn or garden, providing the water chlorine residual level is not above 3 parts per million (ppm) (see Chap. 7). Chlorine levels higher than that may "burn" the grass!

Some rotary backwash valves also include a *rinse* feature, so that after the water has backwashed through the filter, clean water from the pump can be directed to clean out the pipes before returning to normal filtration. This prevents dirt in the lines from going back to the spa after backwashing.

Do not rotate any backwash valve while the pump is running, or you might damage the valve and the pump/motor.

Makes and Models

Now that we have a brief introduction to typical spa filter types, how do we make a selection and what size do we need for a given job? The intended use of the spa and/or the local building and health codes will guide us in answering these questions.

Sizing and Selection

Given that there are at least three types of filtration, how do you select the one that's right for your spa? First, you need to know the required size; then you can apply several simple selection criteria.

SIZING

In many jurisdictions, the required turnover rate for a spa is once every 30 minutes. Using this information and knowing the total volume of your spa, you will know how many gallons or liters per minute the filter must be able to handle.

As an example, my spa is 1000 gallons (3785 liters). To turn that over every 6 hours, I need a filter that can handle

1000 gallons (3785 liters) in 30 minutes

$$= 1000 \text{ gallons} \div 30 \text{ minutes (or } 3785 \text{ liters} \div 30 \text{ minutes)}$$

$$= 33.3 \text{ gallons per minute (or } 126 \text{ liters per minute)}$$

Many building or health codes also determine the maximum number of gallons or liters per minute of water flow permitted for every square foot or square meter of a filter's surface area. Just for the record, many codes also specify the minimum rate of flow during backwash.

These may differ for residential and commercial pools, but many jurisdictions require the following:

Filter style	Maximum flow, gpm/square foot (liters/1000 square centimeters)		Minimum backwash flow, gpm/square foot
High-rate sand	15	(57.0)	15
DE	2	(7.6)	2
Cartridge	0.375	(1.5)	No backwash

Note that cartridge filters are rated at 1 gpm per square foot (3.8 liters per 1000 square centimeters) maximum flow rate on most residential applications and the more stringent 0.375 gpm per square foot on commercial installations. This is because cartridge filters are used mostly on spas where many bathers sit in a relatively small amount of water, creating lots of bacteria, oil, and dirt for the filter to handle. The 0.375 rule simply means a commercial installation will have a larger filter than one at home where the bather load will probably be less.

So with this information, if we chose a sand filter, we divide the required flow rate in gpm for our sample spa, 33, by the maximum flow rate of our local code for sand filters, 15. The next step is to divide 33 by 15, which equals 2.2 square feet. This spa therefore needs a sand filter of just over 2 square feet. Since sand filters are not made this small, a spa of 1000 gallons would not be serviced by a sand filter unless it were attached to a pool with a larger volume of water.

Using the same example, my 33 gpm rate with a DE filter is

$$33 \div 2 = 4.1 \text{ square feet}$$

So we need a DE filter of at least 4 square feet. There are no DE filters this small either, so again this filter medium would not be used for a 1000-gallon spa by itself.

Finally, we have the cartridge filter in the same application:

$$33 \div 0.375 = 88 \text{ square feet}$$

There are plenty of cartridge filters of this size. Just for comparison purposes, note that we would need an 88-square-foot cartridge filter to do the same job as the 2-square-foot sand filter or the 4-square-foot DE filter.

Another sizing and selection criterion is dirt. How dirty does the body of water get that this filter must service? You want to oversize the filter a bit so you don't have to clean it as often. This time between filter cleanings is called the *filter run*. So you might as well get a filter 20%, 50%, or 100% larger than you actually need so it can hold more dirt and thereby leave more time between cleanings—right?

Well, not exactly. If the pump cannot deliver the flow rate in gpm for the square footage of this huge filter you have just installed, then only part of the tank will fill with water, effectively giving you a smaller filter anyway. Also, the pump won't backwash the filter completely if it can't match the gpm rating, which is why plumbing codes call for a minimum backwash gpm flow rate. Make sure your pump is rated for a flow rate at least the same as or greater than that of the filter.

SELECTION

Knowing the size filter you need for your spa is the first selection criterion. Price, size, and efficiency might be additional criteria to consider.

Price is a factor of the current market and supplier availability in your area. Check with your spa supply store. The overall size of the units in the above example might be a deciding factor if the equipment area is small. Sand filters are large, even in the lowest square footage configuration. A cartridge filter is much more compact, but the filtration medium (the cartridge) can be expensive. A DE filter will be slightly larger than a cartridge filter, but the cost of DE is less than that of sand or cartridges.

Finally the efficiency of the filter is worth considering. Filters are rated based on the size of particle they can effectively remove from the water as it passes through the medium, expressed in microns. A micron is a unit of measurement equal to one-millionth of a meter, or 0.0000394 inch! To put it another way, the human eye can detect objects as small as 35 microns, talcum powder granules are about 8 microns, and table salt is about 100 microns. However, here's a rule of thumb:

■ Sand filters strain particles down to about 60 microns

■ Cartridge filters strain particles down to about 20 microns

■ DE filters strain particles down to about 7 microns

The efficiency of DE filters is affected by the cleanliness of the medium. Remember that DE is composed of millions of tiny spongelike particles. As the DE becomes clogged with fine particles, it will lose the ability to strain the smallest impurities. Sand becomes less efficient as it ages. The sharp edges of the grains become round, allowing the finer dirt to pass by.

Because cartridges do not rely on organic material such as sand or DE, they are not affected the same way—as long as the cartridge surface area remains clean, it will filter the same size particle. However, extremely old cartridges that have been acid-washed many times will stretch out somewhat, creating a mesh that is not so fine as when new. Dirt, however, does affect the cartridge filter, as it affects all filters, but with the cartridge it may actually have a positive effect!

Dirt makes the mesh of the polyester cartridge material even tighter, essentially acting as DE. I have known some service technicians who even add a small handful of DE to a cartridge filter after cleaning it, to start this process. Thus, with each successive pass of water through a cartridge filter, it will become *more* efficient as the mesh gets finer and finer from the retained particles. This is also true with sand filters up to a point. Of course at some point, dirt prevents water from passing through the filter and the unit ceases to function properly.

Repairs and Maintenance

Since there are no moving parts on a filter when it operates and few when it is at rest, there's not much to break down. Therefore, when a filter does have a problem, it is usually easy to repair. Since cartridge filters are the most common in spa applications, we will focus on repair and maintenance of those. If your spa is large or is part of a pool/spa combination, and you have a sand or DE filter, repair and maintenance information can be found in *The Ultimate Pool Maintenance Manual.*

Filter Cleaning and Cartridge Replacement

RATING: EASY

The most important thing you can do for a spa is to keep a clean filter. It is also the simplest way to ensure that the other components work up to their specifications. Cleaning a cartridge filter is perhaps easiest of all (Fig. 4-7).

1. **Cut Off the Electricity** Shut off the pump by switching off the circuit breaker. This will make sure it won't come back on (from the time clock, for example) until you're ready.

2. **Isolate the Filter Plumbing** Most spas will have valves that allow you to close off the water from the equipment. Close valves on both the suction and discharge sides of the plumbing. If you don't have isolation valves and the filter is below the level of the water in the spa, drain the filter by opening the drain plug or backwash valve.

3. **Disassemble the Filter** Remove the retaining band (Fig. 4-7A) and lift the filter tank or lid from the base (Fig. 4-7B). Remove the cartridge.

4. **Clean (or Replace) the Cartridge** Light debris can simply be hosed off (Fig. 4-7C), but examine inside the pleats of the cartridge (Fig. 4-7D). Dirt and oil have a way of accumulating in there. Never acid-wash a cartridge. Acid alone may cause organic material to harden in the web of the fabric, effectively making it impervious to water. Soak the cartridge in a garbage can of water with trisodium phosphate (1 cup per 5 gallons, or 250 milliliters per 10 liters) and muriatic acid (1 cup per 5 gallons, or 250 milliliters per 10 liters). About 1 hour should do it. Remove the cartridge and scrub it clean in fresh water. *Don't use soap.* No matter how well you rinse, some residue will remain, and you will end up with suds in your hot tub!

5. **Reassemble and Restart** Reassemble the filter in the reverse order from which you disassembled it. Open the valves to allow water to flow back into the filter. Check for leaks around the lid clamp. If any appear, go back over steps 2 through 5 and be sure the O-rings

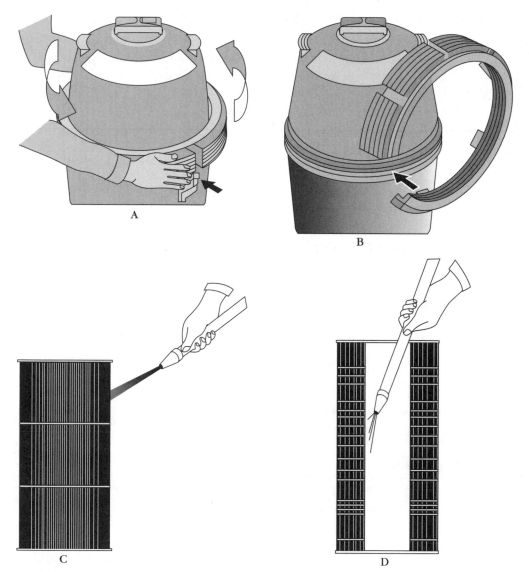

FIGURE 4-7 Cleaning a cartridge filter. *Sta-Rite Industries.*

or gaskets are properly seated and free of debris and that the clamp is tightened adequately. If leaks persist, review the leak repair procedures described below. When everything is reassembled and free of leaks, restart the pump and bleed the air from the filter through the air relief valve. Add more water to the spa if you have partially drained it during the cleaning process.

It's a good idea is to buy a spare cartridge so you can install it and soak the dirty one at your convenience. Replace cartridges when they just won't come clean, when the webbing of the fabric appears shiny and "closed," or when the fabric has begun to deteriorate or tear.

Replacing a Cartridge Filter Unit

RATING: ADVANCED

Replacing a cartridge filter unit is one of the simpler plumbing jobs you can do, since there is no electricity or gas to hook up and normally you will be dealing with no more than three pipes, using techniques described in Chap. 2.

1. **Cut Off the Electricity** Shut off the pump by switching off the circuit breaker. This will make sure it won't come back on (from the time clock, for example) until you're ready. Since PVC plumbing connections should be allowed to dry overnight, this can be a real problem, even causing the spa to drain out completely, so the simplest way to be sure is to shut off the electrical supply at the breaker.

2. **Isolate the Filter Plumbing** Most spas will have valves that allow you to close off the water from the equipment. Close the valves on both the suction and discharge sides of the plumbing. Drain the old filter tank by opening the drain plug or backwash valve.

3. **Cut the Plumbing** The equipment on many spas today is plumbed using unions, making removal of the old filter easy. In this case, you simply unscrew the unions on each pipe and remove the unit. If there are no unions, cut the pipe between the pump and filter in a location that makes connecting the new plumbing easiest. There's no rule of thumb here, just common sense. If the original installation has more bends and turns in the pipe than needed, now is a good time to cut all that out and start over. Eliminating unnecessary elbows

---

TOOLS OF THE TRADE: FILTERS

- Heavy, flat-blade screwdriver
- Hacksaw
- PVC glue
- PVC primer
- Pipe wrench
- Teflon tape
- Silicone lube
- Needle-nose pliers
- Hammer
- Emery cloth or fine sandpaper
- Vise-Grips

will increase flow and reduce system pressure. Then cut the pipe between the old filter and the heater, using the same guidelines. Last, cut the waste pipe, if there is one plumbed into a drain.

4. **Remove the Old Filter** Even without water, the old filter may be heavy, so you may want to disassemble it for removal. Unclamp the lid from the base, and take the parts out separately. Save any useful parts. Old, but still working, valves, gauges, air relief valves, lids, lid O-rings, cartridges, and other components make great emergency spare parts.

5. **Assemble the New Filter** After you remove it from the box, make sure all the pieces are there—cartridge, pressure gauge, lid clamp, and air relief valve. Most new filters come with instructions, and it really pays to read these. While the unit is out in the open and easy to access, screw into place the appropriately sized MIP fittings (1½- or 2-inch; 40- or 50-centimeter), after applying a liberal coating of Teflon tape or pipe dope (see Chap. 2). Some manufacturers include their preference for tape or pipe dope right in the box. Most cartridge filters have threaded openings in a plastic base. Be careful not to overtighten—you'll crack the material you're screwing the MIPs into. If your old filter had unions, unscrew the entire assembly from the old filter and screw it into the new one.

6. **Set Up the New Filter** Put the new unit in the location by the pump, and figure out the necessary plumbing between the pump and the filter; between the filter and the heater; and between the filter and the waste (if appropriate). Some creativity and planning here will save many service headaches later. Basically, you want to avoid as many elbows as possible, and you want to leave enough room between the pump, filter, heater, and pipes for service access (Fig. 4-8A). Remember, you'll have to clean this filter someday. Can you easily access the lid? Will water flowing out of the tank as you clean it flood the pump? Can you access the backwash valve and outlet as needed?

7. **Plumb the New Filter** Use the plumbing instructions in Chap. 2, and make careful connections. Someone once said, "There never seems to be enough time to do it right, but there's always enough time to do it over when it leaks!" Avoid excess angles in the plumbing, which can restrict flow rates (Fig. 4-8B).

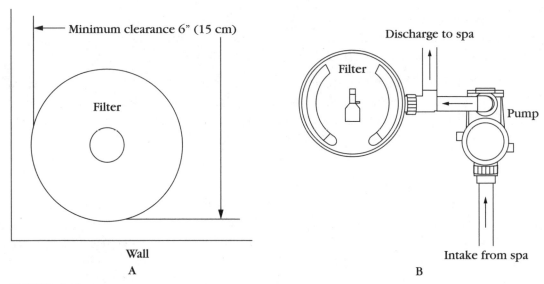

FIGURE 4-8   (A) Placement of a filter. (B) Plumbing a filter.

8. **Start Up** Starting up the newly installed filter is just like restarting after cleaning, as outlined above. Of course, wait for all glue to dry completely, following the instructions on the product container.

Leak Repairs

RATING: EASY

Filters, being rather simple creatures, don't have many repair or maintenance problems beyond cleaning, as discussed. The biggest general complaint, however, relates to leaks of various kinds.

Lids on filters leak in two places: the O-ring that seals them to the tank and/or the pressure gauge air relief valve assembly (Fig. 4-9). The lid O-ring can sometimes be removed, cleaned, turned over (or inside out), and reused.

Some filters will crack on the rim of either the lid or the tank where the O-ring is seated. Obviously, the problem in this case is not a bad O-ring, but a bad lid or tank. Inspect these stress areas carefully for hairline cracks that may be the problem.

Air relief valves sometimes leak if they become dirty or simply worn out. Some are fitted with an external spring that applies tension to create the seal, and when the spring goes, so does the watertight seal. Others have a small O-ring on the tip of the part that actually screws in to create the seal. Unscrew this black plastic type of valve all the way. The "bolt" part will come out to reveal the O-ring on the tip that makes the seal, and you can easily replace that.

The entire air relief valve itself simply unscrews from the T assembly. Apply Teflon or pipe dope to the new one, and screw it back in place.

The pressure gauge also just threads into the T assembly (Fig. 4-9). If you have a leak there, unscrew the gauge, apply Teflon tape or dope to the threads, and screw it back into place. If the gauge doesn't register or seems to register low, take it out and clean out the hole in the bottom of the gauge. Dirt can clog this small hole, preventing water from getting into the gauge for accurate pressure readings.

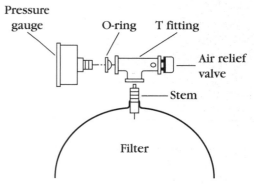

FIGURE 4-9 T fitting with air relief valve and pressure gauge.

For both of these procedures, remember when removing an air relief valve or gauge that you must secure the T with one wrench or pliers while removing the component with the other. The T assembly can easily snap off the filter lid or at the very least come loose if you fail to hold it secure when removing or replacing a valve or gauge. The T assembly itself can come loose and create a leak where the close nipple passes through the hole in the lid. Here you must remove the lid and tighten the nut from the underside of the lid.

Prevention is always the best cure, so examine cartridges, O-rings, and other components carefully each time you break down a filter for cleaning, and always take your time when reassembling. Sloppy reassembly after cleaning is the cause of more leaks than anything else in filters!

Heaters

The heater is arguably the most important component of the equipment system, since hot water is the main reason people invest in a spa or hot tub in the first place. But let's begin with a point about heaters that can create confusion. The spa heater doesn't work as the water heater in the home does. Customers have asked me countless times why the water coming out of the return line isn't hot, like tap water in the home. The assumption is that the pool or spa heater holds a large reservoir of preheated water, as a household water heater does.

Instead, the spa heater warms the water as it passes over copper coils inside the heater unit. The coils are heated by electric current and the resulting water flow out of the heater is typically 10 to 25°F (5 to 10°C) warmer than the water that originally went in. Electric-fueled heaters (Fig. 5-1) can take several hours to heat even a small spa. A cover will help retain heat in spa water regardless of how the water is heated, speeding up the process somewhat.

Larger spas, or those connected to a pool, are typically heated by gas-fueled heaters, but the operational concept is the same as that of an electric-fueled heater. Heat rising from a gas burner (Fig. 5-2) warms a copper heat exchanger. Water passing through the exchanger is heated with each successive pass.

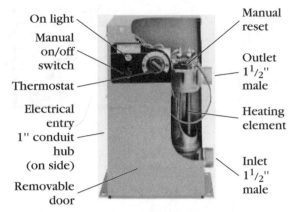

On light

Manual on/off switch

Thermostat

Electrical entry 1" conduit hub (on side)

Removable door

Manual reset

Outlet 1$^1/_2$" male

Heating element

Inlet 1$^1/_2$" male

FIGURE 5-1 Electric-fueled spa heater.

Gas-fueled heaters typically include a sophisticated control circuit, which ensures the proper operation of the unit. Electric-fueled heaters are also designed with the same features, although small electric-fueled heaters (for compact spas) are normally controlled only by a thermostat and the other safety and operational controls are built into the spa control panel.

Finally, a very efficient and cost-effective way to heat a spa is with solar fuel. Much like the gas and electric counterparts, the solar heater works by passing water through pipes that are heated, thus warming the water. With a solar heater, the pipes are laid out inside of panels, which attach to a roof or can be set up on the ground (Fig. 5-3). The sun warms the panels, and as the water passes through the panels, the sun's warmth is transferred to the water. Solar heaters cost no more than their gas or electric counterparts, and once installed, solar heaters cost nothing for the fuel itself.

Let's take a closer look at the components and operation of each type of heater and then examine the ways to keep them operating at peak performance.

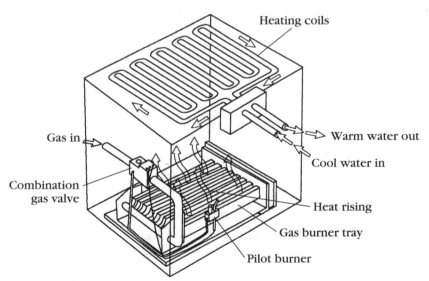

Heating coils

Gas in

Combination gas valve

Warm water out

Cool water in

Heat rising

Gas burner tray

Pilot burner

FIGURE 5-2 Gas-fueled spa heater.

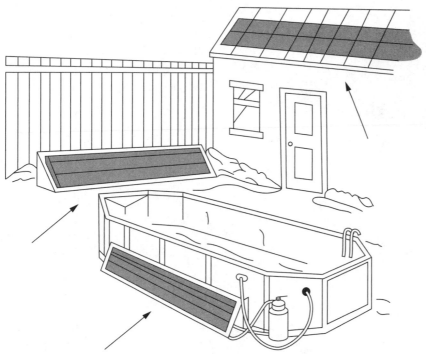

FIGURE 5-3 Solar-fueled spa heater. *Fafco Inc.*

Gas-Fueled Heaters

Figure 5-4 shows the internal components of a typical gas-fueled heater (that which uses natural or propane gas as the heating fuel). The water passes in one port of the water header (item 13) and through the heating coil (heat exchanger), then out the other port.

The heat exchanger tubes are made of copper, which conducts heat very efficiently, so the water picks up 6 to 9°F (3 to 5°C) on each pass. Heat rising from the burner tray (item 1) is effectively transferred to the water in the exchanger because of the excellent conductivity of copper. The tubes have fins to absorb heat even more efficiently and are topped with sheet metal baffles to retain the heat.

The front header has a flow control assembly. This spring-loaded valve is pressure-sensitive, designed to mix cool, incoming water with hot, outgoing water to keep the temperature in the exchanger from becoming excessive. This design keeps the outgoing water no more than 10 to 25°F (5 to 10°C) above the temperature of the incoming

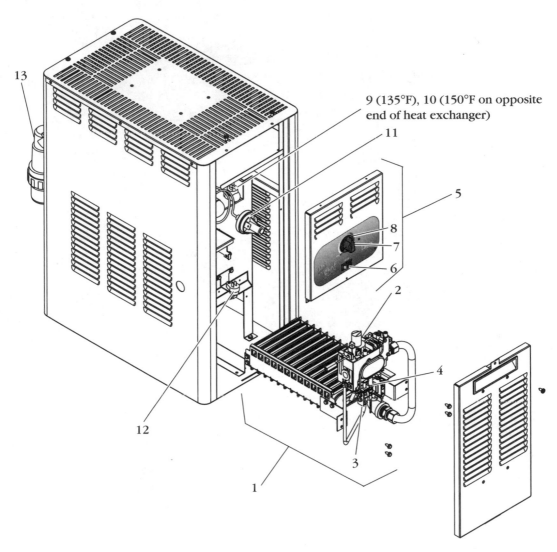

9 (135°F), 10 (150°F on opposite end of heat exchanger)

GAS SYSTEM

1 Burner tray assembly, natural gas
1 Burner tray assembly, LP gas
2 Gas valve, natural gas
2 Gas valve, LP gas
3 Pilot assembly, natural gas
3 Pilot assembly, LP gas
4 Pilot generator

ELECTRICAL SYSTEM

5 Thermostat assembly
6 On/off switch
7 Temperature control knob
8 Stop plate: "Temp-Lok"
9 High-limit switch, 135°F
10 High-limit switch, 150°F
(not shown)
11 Pressure switch, 2 psi
12 Fusible link

WATER SYSTEM

13 I/O header assembly
14 O-ring set, I/O header
(not shown)

OPTIONAL (Not Shown)

15 Propane tank accessory kit
16 External bypass for hot tub/spa
17 Touch-up spray paint: pewter

FIGURE 5-4 Exploded view of typical gas-fueled heater. *Image courtesy of Jandy.*

water to prevent condensation and other problems that greater differentials would create.

The other major component of the gas-fueled heater is the burner tray. This entire assembly can be disconnected from the cabinet and drawn out for maintenance or inspection. Depending on the size of the heater, there will be 6 to 16 burners, the last one with a pilot mounted on it (item 3). Individual burners can be removed for replacement. The combination gas valve (item 2) regulates the flow of gas to the burner tray and pilot and is itself regulated by the control circuit.

Gas-fueled heaters are divided into two categories based on the method of ignition.

The Millivolt/Standing Pilot Heater

As the name implies, the standing pilot system of ignition uses a pilot light that is always burning (Fig. 5-4, item 3). The heat of the pilot is converted to a small amount of electricity (¾ of 1 volt, expressed as 0.750 volt or 750 millivolts) by a thermocouple (item 4) which in turn powers the control circuit. The positive and negative wires of the thermocouple (also called the *pilot generator*) are connected to a circuit board on the main gas valve.

When you light the pilot, it is necessary to hold down the gas control knob to maintain a flow of gas to the pilot itself. When the heat has generated enough electricity (usually a minimum of 200 millivolts), the pilot will remain lighted without you holding the gas control knob down. The positive side of the thermocouple also begins the electrical flow for the control circuit. When electricity has passed through the entire control circuit, the main gas valve opens and floods the burner tray with gas, which is ignited by the pilot.

The Electronic Ignition Heater

When the heater with electronic ignition is turned on, an electronic spark ignites the pilot, which in turn ignites the gas burner tray in the same manner as described above. In all other respects, these types of heaters operate the same as millivolt units.

Regular line current (at 120 or 240 volts) is brought into the heater and connected to a transformer, which reduces the current to 25 volts. This voltage is first routed into an electronic switching device called the *intermittent ignition device* (IID), which acts as a pathway to and from the control circuit. From here the current follows the path through the control circuit switches as described below.

When the circuit is completed, the current returns to the IID, which sends a charge along a special wire to the pilot ignition electrode, creating a spark that ignites the pilot flame. The IID has simultaneously sent current to the gas valve to open the pilot gas line.

When the pilot has lighted, the heat itself generates a current that is sensed by the IID through the pilot ignition wire. This information allows the IID to open the gas line to the burner tray, which is flooded with gas ignited by the pilot.

The Control Circuit

The control circuit (Fig. 5-5) is similar for both millivolt and electronic ignition heaters. The only difference is that one is powered by 25 volts of electricity and the other by less than 1 volt. The control circuit is

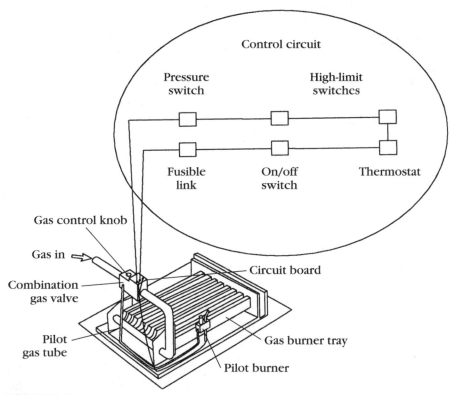

FIGURE 5-5   Heater control circuit.

simply a series of safety switches—devices that test for various conditions in the heater to be correct before allowing the electric current to pass on to the main gas valve and fire up the unit. Following the flow of electricity, a control circuit includes the following parts:

FUSIBLE LINK

The fusible or fuse link (also called a thermal fuse); Fig. 5-4, item 12 is a simple heat-sensitive device located on a ceramic holder near the front of the gas burner tray. If the heat becomes too intense, the link melts and the circuit is broken. This most commonly occurs when debris (such as a rat's nest or leaves) is burning on the tray or when part of a burner has rusted out, causing high flames. Other causes are improper venting (excessive heat builds up in the tray area), extremely windy conditions, or low gas pressure, causing the burner tray flame to "roll out" toward the link.

ON/OFF SWITCH

As the name implies, the on/off switch (Fig. 5-4, item 6) is usually a simple, small rocker-type or toggle-type switch on the face of the heater next to the thermostat control. On older models, the switch may be located on the side of the heater or remotely mounted, so the user can switch the unit on or off from a more convenient location. Manufacturers recommend that a remote on or off switch for a millivolt heater be located no more than 20 to 25 feet (6 to 7 meters) away from the heater. With less than ¾ of 1 volt passing through the control circuit, any loss of power due to heat loss from running along extended wiring means that there may not be enough electricity left to power the gas valve when the circuit is completed.

Also, as the thermocouple wears out and the initial electricity generated decreases, the chance that there won't be enough power becomes very real. Therefore, I suggest from my own experience that remote switches not be located more than 10 feet (3 meters) from the heater and that they be run through heavily insulated wiring to avoid heat loss.

THERMOSTATS

Thermostats (Fig. 5-4, item 7) also called temperature controls, fall into two categories, mechanical and electronic.

The mechanical thermostat is a rheostat dial connected to a metal tube which ends in a slender metal bulb. The tube is filled with oil, and the bulb is inserted in either a wet or dry location, where it can sense the temperature of the water coming out of the heater. These thermostats are not precisely calibrated because so many installation factors will affect the temperature results. In other words, setting the dial at a certain point may result in 95°F (35°C) water for one spa while the exact same setting may result in 105°F (41°C) water for another.

Therefore, spa heater thermostats are generally color-coded around the face of the dial, showing blue at one end for cool and red at the other end for hot (Fig. 5-6). Settings in between are used by trial and error to achieve desired results.

As shipped from the factory, thermostats will not allow water in the spa to exceed 105°F (41°C). Also, they do not generally register water cooler than 60°F (15°C), so if the water is cooler than that, you may turn the thermostat all the way down and the heater will continue to burn. Therefore, the only way to be sure a heater is off is to use the on/off switch.

The electronic thermostat uses an electronic temperature sensor that feeds information to a solid-state control board. These are more precise than mechanical types; however, due to the same factors noted above, they are also not given specific temperatures, but rather the cool to hot, blue to red, graduated dials for settings. Some manufacturers of spa controls make specifically calibrated digital thermostats, but my experience is that no matter what the readout says, the actual temperature may vary greatly. More information about digital and other electronic controls is presented in Chap. 6.

Some heaters are equipped with dual thermostats, so you can set one for a comfortable pool temperature and the other for the spa.

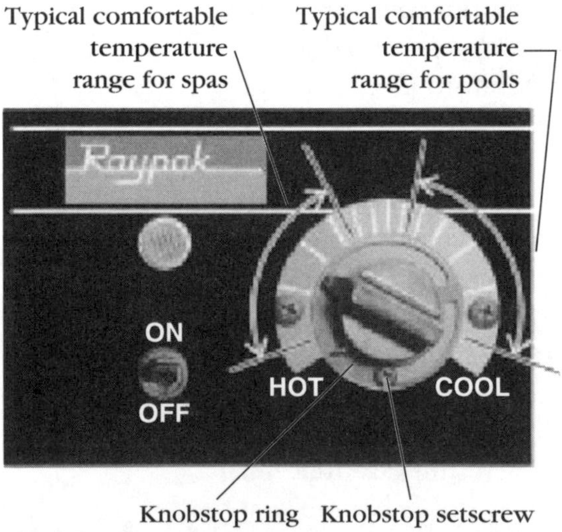

HIGH-LIMIT SWITCHES

While the fusible link detects excessive air temperatures, the high-limit switch (Fig. 5-4, items 9 and 10) detects excessive water temperatures. High-limit switches are small, bimetal switches designed to maintain a con-

Typical comfortable temperature range for spas

Typical comfortable temperature range for pools

ON
OFF
HOT   COOL

Knobstop ring   Knobstop setscrew

FIGURE 5-6   Close-up of thermostat dial. *Raypak, Inc.*

nection in the circuit as long as their temperature does not exceed a predesigned limit, usually 120 to 150°F (49 to 65°C).

Often the first high-limit switch is a 120°F and the other a 150°F switch. They are mounted in dry wells in the heat exchanger header for this reason. Sometimes a third switch, called the *redundant high-limit switch,* is mounted on the opposite side of the heat exchanger for added safety.

PRESSURE SWITCH

The pressure switch (Fig. 5-4, item 11) is a simple switching device at the end of a hollow metal tube (siphon loop). The tube is connected to the header so that water will flow to the switch. If there is not adequate water flow in the header, there will not be enough resulting pressure to close the switch. Thus, the circuit will be broken, and the heater will shut down.

Although preset by the factory (usually for 2 pounds per square inch, or 1 millibar), most pressure switches can be adjusted to compensate for abnormal pressures caused by the heater being located unusually high above or below the water level of the spa.

AUTOMATIC GAS VALVE

Often this is called the combination gas valve (Fig. 5-4, item 2) because it combines a separately activated pilot gas valve with a main burner tray gas valve (and sometimes a separate pilot-lighting gas line combined with the pilot gas valve).

After the circuit is complete, the electricity activates the main gas valve which opens, flooding the burner tray. The gas is ignited by the pilot, and the heater burns until the control circuit is broken at any point (Fig. 5-5), for example, when the desired temperature is reached and the thermostat switch opens; when the fuse link detects abnormal heat and breaks the circuit; when the on/off switch is turned to off; or when the pressure drops (when the time clock turns off the pump/motor) and the pressure switch opens and breaks the circuit.

Figure 5-7 shows a close-up of a typical combination gas valve. The control circuit is connected to the terminals on the valve. The gas plumbing of the automatic gas valve is self-explanatory. The large opening (½- or ¾-inch; 13- or 19-millimeter) on one end, with an embossed arrow pointing inward, is the gas supply from the meter. Note that it has a small screen to filter out impurities in the gas, such as rust

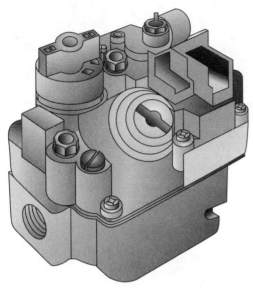

FIGURE 5-7 Typical combination gas valve.

flakes from the pipe. The hole on the opposite end feeds gas to the main burner. The small threaded opening is for the pilot tube, and a similar hole is for testing the gas pressure. These are clearly marked.

Automatic gas valves are also clearly marked with their electrical specifications, model numbers, and, most important, "Natural gas" or "Propane." Black components or markings usually indicate propane.

All combination gas valves have on/off knobs. On 25-volt units, the knob is only on or off. With standing pilot units, there is an added position "pilot" for when you are lighting the pilot. As a positive safety measure in most, you are required to push the knob down while turning.

Natural versus Propane Gas

The differences between heaters using natural gas and those using propane gas are nominal. Most manufacturers make propane heaters in standing pilot/millivolt models only. Because of different operating pressures, the gas valve is slightly different (although it looks the same as a natural gas model), as are the pilot light and the burner tray orifices. The gas valve is clearly labeled "Propane." The heater case, control circuit, and heat exchanger are all the same as those for a natural gas model.

Although there is a section later in this chapter regarding heater safety, it is essential to mention the danger of working with propane right here. Natural gas is lighter than air and will thus dissipate somewhat if the burner tray is flooded with gas but not ignited for some reason. Similarly, the odor perfume added to natural gas will be detected if you are working nearby as the gas floats out and upward. Make no mistake, this is still a serious situation, and explosions can occur.

With propane, however, the gas is heavier than air, and as it floods the burner tray without being ignited, it tends to sit on the bottom of the heater. Because it remains undissipated and because you are less likely to smell it since it is not floating out and upward, if it does suddenly ignite, it will do so with violent, explosive force. Rarely is the heater itself damaged—the explosion takes the line of least resistance, which is out through the open front door panel and into your face, which is probably poised in front of the opening trying to learn why the heater hasn't yet fired! There is more on safety procedures later, but for now just remember—treat propane with great respect.

Electric-Fueled Heaters

Electric-fueled heaters, like the one in Fig. 5-1, are built as stand-alone units with a similar appearance and control circuit as their gas-fueled counterparts. On small portable spas, the heater is mounted to other equipment in a skid pack (described in Chap. 6) with controls that are integrated into the solid-state control panel which manages all the other spa equipment.

Figure 5-8 shows an exploded view of a typical electric-fueled heater. Water flows into the element tank (item 3) and is heated by the element (item 1). As you can see, this heater is served by the same control circuit components as gas-fueled models are.

The smaller versions of electric-fueled heaters employ the same components, but without a separate cabinet for the heater. The element tank with the electric heating element will be mounted as part of the system plumbing, directly after the pump. As described throughout the book, the water usually flows from the pump to a filter and then to the heater; but in portable spas, the filter is located in the skimmer, right inside the spa itself, so the water entering the pump has already been cleansed.

Heater Selection and Sizing

Each manufacturer makes different claims about the efficiency and effectiveness of its heaters, but all estimates depend on a spa that is covered and generally isolated from severe weather elements, such as wind, snow, or extremely cold temperatures.

Gas heaters are rated by the output of heat as expressed in British thermal units (Btu's). Gas heaters for spas range from 50,000 to 200,000 Btu. Electric heaters are rated by the kilowatts consumed and, therefore, the Btu's produced. Here is a comparison of energy use and output of the most common models:

- 1.5-kilowatt (1500-watt) heater = 5119 Btu
- 5.5-kilowatt (5500-watt) heater = 18,750 Btu
- 11.5-kilowatt (11,500-watt) heater = 37,500 Btu

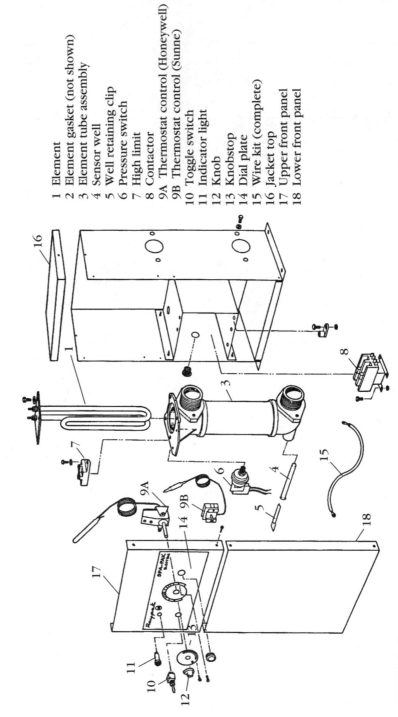

1 Element
2 Element gasket (not shown)
3 Element tube assembly
4 Sensor well
5 Well retaining clip
6 Pressure switch
7 High limit
8 Contactor
9A Thermostat control (Honeywell)
9B Thermostat control (Sunne)
10 Toggle switch
11 Indicator light
12 Knob
13 Knobstop
14 Dial plate
15 Wire kit (complete)
16 Jacket top
17 Upper front panel
18 Lower front panel

FIGURE 5-8 Exploded view of electric-fueled heater. *Raypak, Inc.*

The following chart compares typical electric-fueled and gas-fueled heaters in terms of performance:

Temperature Increase (per Hour) in Degrees Fahrenheit

	500 gallons (1892 liters)	1000 gallons (3785 liters)
1.5-kilowatt electric	2°	NA
5.5-kilowatt electric	5°	2°
11.5-kilowatt electric	10°	5°
50,000-Btu gas	10°	4°
100,000-Btu gas	20°	10°
150,000-Btu gas	28°	14°
200,000-Btu gas	35°	18°

As you can see, electric heaters are not practical in larger spas. Another way to calculate the thermal output of an electric heater is with this formula:

Heater input (kilowatts) $\times$ 410 ÷ spa total gallons

= degree (Fahrenheit) increase per hour

Using this formula, we assume a 500-gallon spa and a 5.5-kilowatt heater. The calculation looks like this:

5.5 (kilowatts) $\times$ 410 = 2255

2255 ÷ 500 gallons = 4.5°F increase per hour

Cost of Operation

The cost of operating a heater is simple to figure out if you know what your customer pays for 1 therm of gas or 1 kilowatt of electricity.

A therm—the unit of measurement on your gas bill—is 100,000 Btu per hour of heat. My last gas bill showed I pay about 50 cents per therm. The heater model tells you how many Btu's per hour your heater uses. Divide that number by 100,000 to find out how many

TRICKS OF THE TRADE: HOW LONG WILL IT TAKE TO HEAT A SPA?

Another way to estimate how long it takes to heat your spa is to use a table like this one. Find the volume closest to that of your spa in the left-hand column, and find your heater in the row across the top.

	Heater size				
Spa size	125,000 Btu	175,000 Btu	250,000 Btu	325,000 Btu	400,000 Btu
200 gallons	30 minutes*	20	15	12	<10
400 gallons	60	45	30	25	20
600 gallons	90	65	45	35	30
800 gallons	120	85	60	45	40
1000 gallons	150	110	75	60	47

100 gallons = 378 liters.
*Minutes required for every 30°F temperature rise desired.

therms per hour that will be. Next, determine how many hours of operation are needed to bring the temperature up to the desired level. Let's try an example with a 200,000-Btu heater operating for 2 hours each day:

$$200{,}000 \text{ Btu} \div 100{,}000 = 2 \text{ therms per hour}$$

$$2 \text{ therms} \times 2 \text{ hours} = 4 \text{ therms per day}$$

$$2 \text{ therms} \times 50 \text{ cents} = \$1 \text{ per day}$$

Of course this calculation assumes the heater is operating for the entire 2-hour period. If you reach the desired temperature in less time, the heater will shut off and will restart only as needed to maintain that temperature.

How much does it cost to keep a standing pilot burning in a millivolt heater? Pilots use between 1200 and 1800 Btu per hour, so the cost calculation is the same as the formula already described. By the way, the temperature of that little flame is over 1100°F, so when you remove

---

TRICKS OF THE TRADE: MAKE MY SPA HOT...NOW!

Sometimes everything is working fine, but the spa user just wants it to heat up faster. Here are a few tips to speed up the process:

- Double-cover the spa to keep it warmer when it is not in use. Add a floating cover (or two) on the water surface and an insulated cover over the entire unit.
- Insulate the spa itself. If it is above-ground, consider dropping the entire unit into the ground and insulate with sand around the sides.
- Reset the timer. Find out when the user most often wants the spa hot, and set the pump and heater to operate automatically for 1 hour before that time.
- Insulate plumbing between the equipment and the spa.
- Upgrade the heater or add a second one. Yes, there's nothing wrong with putting a second heater in line after the first one. Each will raise the water temperature 6 to 10°F on each pass.
- Use solar heat (see Chap. 6). Since the spa probably contains less than 1000 gallons (3785 liters), you won't need many panels for effective solar heating, which you can operate daily as the spa filters for efficient, no-cost heating.

---

a pilot assembly for repair, don't grab one that has recently been lighted.

Electricity is sold by the kilowatt-hour (1000 kilowatts consumed per hour) which equals 3412 Btu. I pay 15 cents per kilowatt-hour, so if I have a 5.5-kilowatt heater, the calculation looks like this:

5.5 kilowatt $\times$ 15 cents per kilowatt-hour = 82.5 cents per hour

Once again, the actual daily cost will depend on the desired temperature and starting temperature of the water.

Heater Installation and Repair

Installing a new heater or repairing an existing unit may seem complicated, but it is actually easy to manage if you follow the path of the water and fuel (gas or electricity). Since the heater combines water, electricity, and gas, it is important to use extra care when you are working around these units, following all manufacturer's warnings and procedures.

Installation

RATING: PRO

This section deals with installing a stand-alone heater regardless of the fuel source. In Chap. 6, the "Skid Packs" section describes the installation and servicing of integrated heater units for compact equipment packages in portable spas. Any heater installation is composed of four basic steps—location (including ventilation), plumbing, gas connections, and electrical connections.

Although installing a heater uses simple skills described throughout this book, it is rated Pro because any mistakes can be disastrous. No matter what the fuel, improper installation of a heater can result in fires, explosions, and injuries. If you feel ready to tackle the job, exercise extreme caution with each step and follow the manufacturer's guidelines and installation instructions. The following steps are meant as general guidelines only.

1. **Location** Hot air rises. Very hot air rises very quickly in large volumes, requiring replacement by adjacent cooler air. Burning fossil fuel, such as gas, results in by-products such as carbon monoxide, which is deadly. These simple concepts are at the heart of the decisions about where to locate your heater.

   Most residential heaters are designed to be installed along with the pump/motor and filter at an outdoor or indoor location. When purchasing the heater, you ask for a *stackless* heater (Fig. 5-4 shows a stackless top) for an outdoor installation, meaning there will be a draft hood on top of the heater, but no additional vent pipe or "stack" to remove excess heat and the products of combustion (carbon monoxide). If your installation is to be indoors, ask for the *stack* heater, which comes with a vent hood for attachment to a stack pipe.

   Never use a "stackless" top on an indoor installation. Burning gas produces carbon monoxide, and even if the heater is small and the indoor location large and well ventilated, this is a deadly gas. Of course electric-fueled heaters do not have this problem, and the exterior case does not heat up.

   Take the heater out of the box and look for instructions. *Read them*! I have installed over 500 heaters in my days, but constantly changing designs mean I still read those instructions before working with any new unit.

Open the front panel on the heater. Inside you will find a plastic bag with installation hardware (the owner's manual may also be packed in here). Set the heater near the filter, but remember you will need to do future service work on both the filter and the heater, so leave enough space for moving around. The other spacing guidelines are provided by the American National Standards Institute (ANSI) and are devised to keep carbon monoxide from entering closed living spaces of your home and to allow enough draft area around the heater. Figure 5-9A and B shows typical clearances for gas heaters; electric heaters simply require 6-inch (15-centimeter) clearance on all sides and 18 inches (45 centimeters) on top and in front for maintenance access.

Set the heater on a solid, level, noncombustible base. Although heat rises, the metal cabinet (of gas heaters) will get quite hot underneath as well. Heaters have their own "feet" or runners to hold them off the surface; however, even these get warm and would not be compatible with, say, a wooden floor or carpet. The best flooring is a concrete pad.

> ### TOOLS OF THE TRADE: HEATERS
>
> - Flat-blade screwdriver
> - Phillips-head screwdriver
> - Hacksaw
> - PVC glue
> - PVC primer
> - Pipe wrench
> - Teflon tape
> - Silicone lube
> - Needle-nose pliers
> - Hammer
> - Emery cloth or fine sandpaper
> - Pipe wrench
> - Multimeter (and millivolt tester if separate)
> - Nut driver set
> - Electric drill with reaming brush
> - Channel lock–type pliers
> - Kneepad

2. **Venting** As noted, outdoor installations require no additional venting; they are vented sufficiently within their stackless tops. In windy areas, however, you may want to consult the manufacturer's recommendations and add a short stack and cap (about 3 feet total) to cut down on excess drafting. Refer to each manufacturer's specifications on this.

For indoor installations (Fig. 5-9B), follow the guidelines in the heater booklet. Working with stack pipe made of flexible sheet aluminum, which you get at your supply house, is no tougher than working with PVC, except there is no glue—they just snap together.

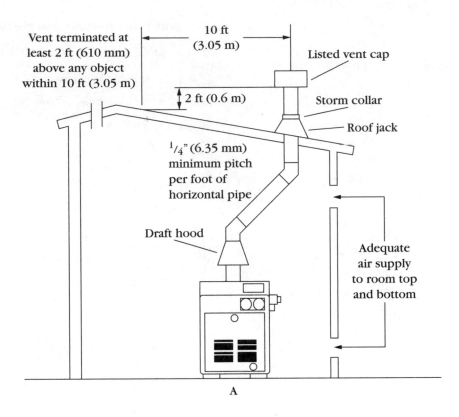

Vent terminated at
least 2 ft (610 mm)
above any object
within 10 ft (3.05 m)

10 ft
(3.05 m)

Listed vent cap

2 ft (0.6 m)

Storm collar

Roof jack

$^1/_4$" (6.35 mm)
minimum pitch
per foot of
horizontal pipe

Draft hood

Adequate
air supply
to room top
and bottom

A

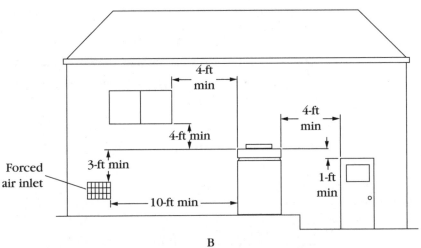

4-ft
min

4-ft
min

4-ft min

Forced
air inlet

3-ft min

1-ft
min

10-ft min

B

FIGURE 5-9 (A, B) Location of a gas heater. (C) Plumbing a heater. *A: Image courtesy
of Jandy. B: Raypak, Inc.*

Vent pipe comes in lengths of 2 to 6 feet (60 to 180 centimeters) and diameters of 3 to 12 inches (7 to 30 centimeters). You also buy preformed 45- and 90-degree elbows and connector sleeves just as with PVC plumbing.

Going through walls or a roof to get outside, however, is trickier, since you need to use double-wall pipe, flashing, and waterproofing to make the passage safe and rainproof. Some local codes require a permit and a licensed building contractor to handle this part. If your indoor installation is a replacement of a heater that otherwise worked well (and

C

FIGURE 5-9 *(Continued)*

your replacement has the same Btu output), you really only need to connect your new heater venting to the old vent on up and out of the building.

But if it is a new vent stack that means cutting walls and a roof, I'd hire a roofing contractor to handle that part of it. It is well worth the small added cost.

In either case, it is your responsibility to make sure the clearances meet the ANSI code so the heater works properly. Familiarize yourself with the guidelines in Fig. 5-9, and supervise the contractor's work. One simple way to perform a rough test of the venting is to light a match and hold it under the draft hood—the smoke should be drawn up into the vent system and out of the building.

Some howling, whistling, or other ventilation harmonic noise is normal and acceptable. As mentioned, a very large volume of air is rushing up through the heater, and as it passes over vent fins, it might "howl." Notice what is normal for the heater, and then compare that to any future noises. Changes in the sound, however, may denote problems.

A few minutes after the heater fires, a knocking noise which may actually rock the heater is *not* normal. This denotes overheating for some reason and is caused by the superheated water expanding and trying to escape. Some high-pitched whining is caused by debris in the gas line. That line should be disassembled, cleaned, and reassembled.

Electric heaters do not vent significant heat, so there will be no drafting or noise.

3. **Plumbing Connections** No other aspect of installation could be easier, as there is one pipe in and one pipe out. The manifold (the end where the plumbing enters the heat exchanger) is set up for 1½- or 2-inch (40- or 50-millimeter) pipe. Most spa heaters will use simple threaded fittings.

Be sure to check the manufacturer's recommendation regarding the type of pipe that can be plumbed directly to the heater. Some are designed to accommodate ordinary PVC, but others will require the greater heat tolerance of CPVC, a special variety of PVC pipe. Not all local codes permit use of CPVC for this purpose, and not all heaters are designed to allow it either. So check with your local building code and manufacturer's requirements before using it.

CPVC will glue together with ordinary PVC, but check the manufacturer's recommendation for how long the CPVC plumbing needs to be before switching over to the less expensive PVC pipe.

4. **Gas Connections** In replacing a heater, you will generally hook the new heater up to the existing gas line. If you are making a new installation, be careful to note that gas plumbing must follow rigid guidelines, because obviously a gas leak is a greater hazard than a water leak. Given that, the procedures are the same as those described in Chap. 2 and require no special skills or tools, just lots of care.

Each manufacturer will specify the required size of pipe to ensure an adequate supply of gas, depending on the size of the heater and the distance from the source of the gas (natural gas meter or propane tank). Generally, these will be ¾- to 1½-inch-diameter (19- to 40-millimeter) pipes.

Gas plumbing may be steel or special PVC pipe. Some building codes permit special gas PVC pipe for underground runs (a heavy-duty PVC pipe material which is green-tinted to denote gas). It can be argued that especially with underground pipe, PVC might be better because it won't rust. The disadvantage is that anyone digging in the yard can more easily rupture a PVC pipe than a metal one. If you do run PVC gas pipe, it must be accompanied in the ground by a 16-gauge tracer wire, so the gas line can later be found by a metal detector, if needed. Most codes also require that PVC gas lines be buried at least 18 inches (45 centimeters) below ground [12 inches (30 centimeters) under concrete].

Finally, any risers (lines coming up out the ground) or lines above-ground must be metal anyway, so you must transfer from the PVC to metal before leaving the trench. Metal gas pipe must be 12 inches (30 centimeters) below the ground or 6 inches (15 centimeters) below concrete. Metal gas pipe is painted green (to denote gas), and any underground portions must be wrapped with waterproof 10-mil-thick gas tape to at least 6 inches (15 centimeters) above the ground. Lines should be run as close as possible to the meter at one end and to the heater at the other. Most building codes prohibit the use of flexible pipes for gas lines.

Gas hookups must include a shut-off valve just before the heater, followed by a sediment trap (just a drop of pipe off a T that can collect impurities in the gas before it gets to the heater), followed by a threaded union (Fig. 5-10). When you need to service the heater, you can close the valve, break the union, and remove any parts of the system without shutting off the gas to the entire house. Similarly, if a leak or other emergency problem develops in the heater, you can quickly shut off the supply of gas in a convenient location.

If this is a new installation, be sure to check the ability of the gas meter to supply the required amount of gas, especially if your house is simultaneously using a gas clothes dryer, gas water heater, gas stove, or other gas appliances. If in doubt, check with your local gas supplier.

Gas lines must be pressure-tested. Building codes determine how much pressure the line must hold, and special gauges are available at plumbing supply stores. However, I suggest hiring a licensed plumber to conduct this test for you. He or she has the tools and knowledge to conduct this important safety check. When putting a heater into service, however, whether a new installation or just replacing a heater to an existing gas supply line, you can conduct a simple test of your own.

Turn on the gas supply to the heater, and wipe liquid soap over any joints or unions

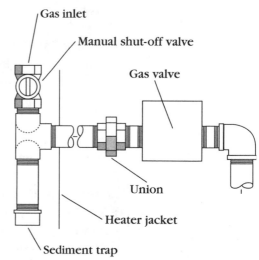

FIGURE 5-10 Gas connections for spa heater.

(ammonia in a spray bottle will also work well). Escaping gas will make the soap or ammonia bubble, so you can detect any leaks quickly. Never leak-test with a match—you might end up in the next county!

5. **Electrical Connections** If the heater has an electronic ignition, you must supply the electricity. Wire should be 14-gauge copper, run in its own waterproof conduit (not shared with wiring for other purposes). For either 120- or 240-volt supply, run three wires. One wire—the ground wire—is green. On a 120-volt supply, the other two wires should be black or red (hot line) and white (neutral). On a 240-volt supply, both wires may be black or red.

Electronic ignition heaters are designed to operate on either 120 or 240 volts with the simple reorganization of the wires that supply power to the transformer. The heater control circuit works on 25 volts regardless of the voltage supplied. Carefully study the wiring diagram provided by the manufacturer to make sure you are connecting 120 volts to 120-volt connections or 240 volts to 240-volt connections.

Where should the electrical supply come from? I prefer to run the wiring from the system time clock, so that the heater can't operate unless power is also being supplied to the pump/motor. This can't guarantee that the heater will come on only when water is flowing, but it is one more safety measure. The ground wire should be attached to the ground lug or bar in the clock and to the ground lug inside the heater cabinet.

As you will see in the "Time Clock" section of Chap. 6, there is a "line" side and a "load" side. Always connect the heater (and pump/motor) to the *load* side; otherwise the wiring to the heater will always be "hot."

Some manufacturers suggest including a "fireman's" switch. This switch is mounted in the time clock and is designed to shut the heater off about 20 minutes before the time clock shuts off the pump. The idea is that by running cool water through the heater for 20 minutes after it has been operating, you will cool the heater components and extend their life. Most time clocks have predrilled holes for standard fireman's switch installation (and directions are provided in the switch package).

Electric-fueled heaters need heavier-gauge wire since they consume large amounts of electricity as the heating fuel. Follow the manufacturer's guidelines for these installations, or bring in an electrician to make the hookup. Because of the load factor, the wiring to the time clock will likely be inadequate to supply the electric heater as well, so plan on running a separate conduit and wires from the breaker box.

Basic Start-up Guide

Whether you have just installed a new heater or are firing up a heater that has not been used in some time, these tips will get you through it safely.

1. **Bleed the Gas Line** If the gas supply line is new, bleed the air out of the pipe by opening the line at the union near the heater and opening the shut-off valve. When you smell gas coming out of the pipe instead of air, shut off the valve and reconnect the union. There is still, obviously, air in the remaining 1 or 2 feet of pipe and in the combination gas valve inside the heater; but this will bleed quickly if gas is present in the line up to the shut-off valve. Open the gas shut-off valve.

2. **Leak-Check** Make sure the heater on/off switch is at *off*. Turn on the pump/motor and make sure the air is out of the water system. Check for leaks!

3. **Fire Up the Heater** If the heater has a standing pilot, follow the instructions in the adjacent text box to light the pilot, or look for instructions provided with the heater. It may take 60 to 90 seconds to get a flow of gas in the pilot tube the first time, as the air bleeds out and is replaced by gas. Never try to light an electronic ignition pilot with a match or other fire source. After you light the pilot, turn on the on/off switch and turn up the thermostat as needed to fire the heater.

   If the heater has an electronic ignition, turn the valve on the combination gas valve inside the heater to *on*. Turn on the on/off switch, and turn up the thermostat as needed to fire up the heater. It may take a minute or two the first time to bleed air out of the system and replace it with gas, so don't worry if it seems to take awhile for the pilot to ignite and the unit to fire the first time.

4. **Check for Normal Operation** When the heater first fires, the heat will burn off the oil that is usually applied by the factory to the heat exchanger as rust prevention. Light smoke for a few minutes is normal. Also normal is some moisture condensation as very cold water runs into the very hot heat exchanger. The condensation will drip down onto the burner tray and sizzle—a little of this is normal, too.

Observe the heater for the first 10 minutes. Make sure the smoke and condensation stop and that there are no leaks. Use your eyes and nose (gas leaks?). Turn the heater off and on a few times to be sure it operates properly. Do this from the on/off switch and the thermostat a few times. Remember, if the water is colder than about 65°F, the thermostat won't shut off the unit because it doesn't register that low. Now shut off the pump. The heater should also shut off within 5 seconds. If it doesn't, the pressure switch needs adjustment (see below).

With the pump running, touch the inlet pipe to the heater and then the outlet. The temperature differential should not be more than 10°F. If it is much more than that, refer to discussion of bypass valves below.

TRICKS OF THE TRADE: TROUBLESHOOTING GUIDE TO HEATER PROBLEMS (SYMPTOMS AND SOLUTIONS)

Spa is not reaching desired temperature.

- **Set the thermostat higher.**
- **Run the circulation system longer and/or set the time clock so that the water heats up before the most common bathing time.**
- **Clean the filter or remove other circulation obstructions.**
- **The heater may be undersized (see the section on sizing).**
- **If burner flame is low ("lazy"), check gas pressure or clear debris from burners.**

Heater makes whining/knocking noise and/or stays on after the pump goes off.

- **Pressure switch is out of adjustment.**
- **Clean the heat exchanger.**

Heater goes on/off frequently.

- **Adjust pressure switch.**
- **Check circulation, clean filter.**
- **Adjust bypass valve (or check that automatic bypass is undamaged).**

Soot forms inside heater.

- **Air supply is restricted—clear.**
- **Flow rate is excessive—correct.**
- **Clear debris from burners.**

Terry's Troubleshooting Tips

This is the only section of the book I name after myself, because after years of troubleshooting heaters, I have developed a method that seems to work for me. I think it will work for you, too. This section will first provide you with my checklist in summary and then explain any tests or repairs that are not self-explanatory in greater detail at the end. Although some of the troubleshooting tips refer to gas-fueled heaters, the majority of the technology and diagnostic tips will apply to any heater.

1. Check the *water* system.
   A. Is the pump primed and running without interruption?
      (1) Enough water in the spa?
      (2) Air purged from system?
      (3) Skimmer/main drain clear?
      (4) Pump strainer pot and impeller clear?
      (5) Filter clean?
      (6) *All* valves open?
   B. Is the water chemistry correct?
      (1) High pH could mean scale in heater.
      (2) Low pH could be causing leaks.
   C. Any visible leaks?
      (1) At exterior plumbing connections.
      (2) At interior heat exchanger components.
2. Check the *gas* system.
   A. Is gas getting to the heater?
      (1) Look for pilot flame.
      (2) Combination gas valve turned to *on?*
      (3) Gas supply lines adequate/unobstructed?
      (4) Propane tank full?
   B. If pilot and/or burner are working…
      (1) Is flame 2 to 4 inches (5 to 10 centimeters) and steady?
      (2) Is flame steady and blue?
      (3) Are all burners fully lighted?
         (A) Does pilot ignite within 5 seconds?
         (B) Does burner ignite within 10 seconds after pilot?
         (C) Does tray ignite without "flash" or loud boom?
   C. Is ventilation adequate?
      (1) Adequate air supply to heater?
      (2) Adequate venting of hot air away from heater?

    D. Smell for leaks!

       (1) Sniff around outside connections and unions.

       (2) Sniff around combination gas valve and joints.

3. Check the *electrical* system.

    A. If electric-fueled heater, is the circuit breaker for the unit on?

       (1) Is the Reset breaker button tripped?

    B. Is the on/off switch on? Is the remote switch (if any) on?

    C. Is the thermostat turned up high enough?

    D. On electronic ignition gas heaters, is 25 volts coming out of the transformer?

       (1) If not, is 120 or 240 volts coming into transformer?

       (2) Is heater grounded properly?

       (3) Are all connections tight and clean?

    E. Check the pilot.

       (1) Is the pilot clear of rust and dampness or insects?

       (2) Has the electronic ignition fired the pilot, and is there a strong, blue flame?

    F. Check the control circuit, following the path of electricity.

       (1) Power from pilot generator or transformer?

       (2) Power to each switch?

          (A) On/off.

          (B) Thermostat.

          (C) Fusible link.

          (D) Pressure switch.

          (E) High-limit (two).

          (F) In-line fuse (some models).

          (G) Fireman's switch.

       (3) Power to gas valve?

Repairs

Now that we have reviewed the list of the most common things that can go wrong, let's turn to more detailed troubleshooting and repair procedures for each of them.

WATER SYSTEM

RATING: EASY

If you can answer yes to all the water flow questions posed in the check-list, yet the flow of water is still too low, the problem may be inside the heat exchanger. Over time and with improper water chemistry, scale

TRICKS OF THE TRADE: GOLDEN RULES OF HEATER REPAIR

- Always turn off the heater when you are making repairs. Preferably turn the pump off as well and disconnect or shut off any source of electricity. If you don't, you may complete the repair or touch some wires together and cause the heater to restart when you don't really want it to. By shutting everything down, you control the entire process—you check your work and control the test-firing when you think you're done and want to check the work. Otherwise, the heater controls you.

- It is generally better to replace components than to repair them. If it's a well-worn heater with other parts that will soon give out, too, replace the entire heater instead of adding new parts every month ad infinitum.

- Heater parts fail for a reason. Find out *why* a part failed in the first place, or it will happen again.

- The majority of heater failures are the result of dirty filters (or obstructed water flow from the spa). In short, the failure has nothing to do with the heater at all. Look around at the entire installation before you start on what has been called in as a "heater problem."

- If the heater has been running prior to any repair you are making, watch out for hot components. Pilots generate over 1100°F, so they stay hot for a long time after they've gone off. Cabinets and other metal parts get hot, too, so watch what you grab.

(calcium "liming") can build up inside the tubes of the heat exchanger, slowly closing the passages, somewhat similar to fat closing the arteries in your body. You can ream out the heat exchanger, but it requires some specialized tools and a willingness to disassemble the entire heater, so you may want to leave it to a service technician. You can, however, prevent it from happening again by taking steps to balance the water chemistry (see Chap. 7).

GAS SYSTEM

RATING: ADVANCED

Check that the heater is being supplied with gas. If this is not apparent (from a working pilot, for example), shut off the gas valve and open the union near the heater. Make sure there are no other open flames or sparks nearby that might ignite the gas; then turn the gas valve back on and listen for gas flow (you will hear the hissing before you will smell the gas). If there is no flow, go to the source—the main meter or propane tank.

TRICKS OF THE TRADE: LIGHTING (OR RELIGHTING) THE PILOT

RATING: EASY

This procedure applies to standing pilot units only since the electronic ignition is automatic unless something is broken. Instructions are almost always printed on the heater itself. Look for these and follow them. If the directions for your particular heater are obscured or missing, the following procedure is most commonly used:

1. **Shut off:** Turn the gas valve control to *off,* and wait 5 minutes for the gas in the burner tray or around the pilot to dissipate (safety consideration). Turn the on/off switch to *off.*

2. **Light the pilot:** Turn the gas valve control to *pilot* and depress. If the area is quiet, you should hear a strong hissing sound as the gas escapes from the end of the pilot. If you do not, it may be clogged by rust or insects (see below for removal instructions). Light the pilot and continue to depress the control. Hold the control down for at least 60 seconds. This allows heat from the pilot to generate electricity in the thermocouple to power the gas valve (which will then electronically hold the pilot gas valve open and power the control circuit, which ultimately opens the main burner gas valve). Release the control, and the pilot should remain lighted.

3. **Verify:** This step is not usually included in instructions printed on the heater—verify that the pilot flame is "healthy." That is, look to see that a strong-burning, blue flame of 2 to 3 inches extends toward the burner tray and that a secondary flame of equal value is heating the thermocouple. Make sure the pilot is still securely in place; it may be burning properly, but rusted fasteners or poor installation from previous repairs may have left it dangling near the burner tray, but not in close contact with it.

4. **Fire the heater:** Turn the gas valve control to *on.* Stand back from the heater (and to one side—not in front of the open door of the heater) in case of *flashback.* The heater should fire normally (remember that the pump must be running for the heater to fire).

REPAIR/REPLACE BURNER TRAY COMPONENTS

RATING: ADVANCED

Rusted out burners or tray components will be obvious to the naked eye. Following Fig. 5-4, remove the burner tray assembly and replace any defective burners (in item 1).

If the pilot fails to light, it may be clogged. If the heater "booms" when it lights, it means the pilot is not igniting the gas in the tray soon enough. Is the flame too small because of an obstruction in the pilot? Is the pilot bent or rusted and not positioned close enough to the burner tray? Take it apart, clean it, and reassemble (or replace it if it is damaged).

There are several types of pilot used in various heaters, but once the tray is removed, the disassembly will be obvious. Let's review typical installations and problems with the standing pilot units and the electronic ignition units.

STANDING PILOT UNITS

RATING: ADVANCED

Disconnect the wires of the pilot generator from the combination gas valve. Remove the pilot gas tube from the combination gas valve, and remove the entire pilot assembly from the tray.

The pilot generator either clips in place or is held by a threaded ring (Fig. 5-11). Remove the pilot generator. If it is rusted or swollen, replace it.

The pilot itself will further disassemble into two or three sections, depending on the model. Examine each part for obstruction or rust. Blow through one end of the pilot gas tube, and be sure it is also unobstructed.

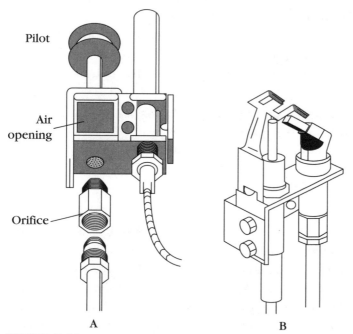

Pilot

Air
opening

Orifice

A   B

FIGURE 5-11   (A) Standing pilot unit. (B) Electronic pilot unit.

The pilot can be cleaned by soaking in muriatic acid for 30 to 60 seconds, cleaning thoroughly with fresh water, and blowing dry (do not leave acid or water on the metal parts, which can cause rust, in or on the pilot).

Natural gas is actually odorless, so to make leaks more detectable, the gas supplier adds a perfume. This perfume has been found to attract insects who build nests in pilot assemblies. Insects and rust are the two main enemies of the pilot.

ELECTRONIC IGNITION PILOT UNITS

RATING: PRO

The ignition electricity feeding the pilot is between 10,000 and 20,000 volts, so be sure the power is off before servicing! Remove the high-tension wire that supplies electricity to the pilot. Remove the pilot gas tube from the combination gas valve, and remove the pilot assembly from the burner tray.

The IID senses the pilot by the electricity generated from the heat of the flame—a "reverse" current is generated along the line of 0.00002 amp (2 milliamps), which is so sensitive that any electrical obstruction such as corrosion prevents this "signal" from reaching the IID. I have cleaned and reassembled these units and successfully put them back into service; however, they usually fail again soon. In this case, replace the pilot assembly, including the wire between the IID and the pilot unit.

As described above with standing pilot units, disassemble the pilot assembly (Fig. 5-11B) and supply tube, and clear any obstructions or rust. Reassemble it the opposite way from how it came apart.

AUTOMATIC COMBINATION GAS VALVES

RATING: ADVANCED

There are no in-the-field repairs that can be made to the automatic combination gas valve. If you determine that it is failing, replace it. The plumbing of the valve is female-threaded and screws directly onto the gas pipe of the burner tray. When replacing it, apply Teflon tape to the male gas pipe of the burner tray. I do not like pipe dope in this case because it can too easily squeeze off the threads and into the opening of the gas valve, obstructing gas flow.

TRICKS OF THE TRADE: HEATER SAFETY

Heaters are unquestionably the most dangerous component of a spa equipment group. They combine water under pressure and heat, gas, or other combustible fuel and electricity. The point is simply that whatever care you exercise normally must be doubled when you are working with heaters. Therefore, I have a simple safety checklist for working around heaters:

• Never bypass a safety control and walk away. Jumping controls is a good way to troubleshoot, but not to operate the unit. Always remove your jumper cables after troubleshooting.

• Never repair a safety control or combination gas valve. Replace it. You will notice that your supply house doesn't even sell parts for gas valves—they should never be repaired, because future failure could be catastrophic.

• Never hit a gas valve—it may come on, but it may stay on!

• Keep wiring away from hot areas and sharp metal edges of the heater.

• When you are testing a heater, especially after a repair, keep your face and body away from the burner tray where flashback might occur.

VENTILATION

RATING: ADVANCED

You would be surprised at how many problems are created by improper ventilation. You need only place your hand about 2 feet over the top of a gas-fueled heater to feel the power and volume of hot air moving through the ventilation system and to understand that any restrictions under such heat will result in damage of some kind.

Sooting—black carbon build-up on the heat exchanger—is the symptom of improper ventilation. Carbon starts as a dirty, black coating and builds up to the point where chunks of "coal" burn and break off, falling down on the burner tray below. The burners becomes clogged, shutting down the heater's full capacity. The heater smokes when it operates.

Here's a perfect example of a problem that is easy to repair, but not worth the trouble if you don't also address the cause. Refer to the "Installation" section earlier in this chapter and to the manufacturer's guidelines about proper ventilation. Correct any problems in this area first.

The remainder is easy—remove the burner tray as previously described, cleaning all the components with soap and water and drying out everything before reinstallation. While the burner tray is out, remove the draft hood and any vent stack to reveal the heat exchanger. Using a stiff bristle brush and soapy water, clean the exchanger. Be sure to get inside the cabinet to clean the underside of the exchanger as well (it may be easier on small heaters to remove the heat exchanger for cleaning). Be careful not to soak electrical components or the firebrick.

After you reassemble the heater, small amounts of carbon that you may have missed will burn away over time, providing that you have adequately corrected the ventilation problem and no new sooting is occurring. Do not use a wire brush for soot removal. It can cause sparks, which may ignite the carbon. Use a stiff natural or plastic brush.

ELECTRICAL SYSTEM

RATING: EASY

On electric-fueled heaters, check the Reset button, a safety circuit breaker. Some are activated by an overheating condition, such as when the unit is allowed to run dry. Some are only sensitive to excess amperage, such as the circuit breakers on your house. After they cool (in a minute or so), they can be pressed back in to reset; however, as with other troubleshooting, you must determine the cause of the circuit break in the first place, which is usually overheating due to low (or no) water flow. Check your circulation before resetting and afterward; observe the unit in operation for several minutes to make sure it won't pop off again a minute after you leave. Of course, sometimes these little breakers simply wear out from old age and need to be replaced, but this is the least common fault.

CONTROL CIRCUIT

RATING: PRO

As you can see from the components in Fig. 5-4, replacing the various control circuit switches as needed is very easy. The on/off switch and thermostat (mechanical or electronic) come off the cabinet with the removal of two screws; the high-limit switches are easily removed from the header; the pressure switch simply unscrews from the end of the water tube feeding it; the fusible link is held by a ceramic holder, which is held in place by a screw or two; and the wires to each switch simply unclip or come off by loosening a screw.

The problem is not with replacing any faulty control circuit switch—the problem lies in determining which switch is faulty . . . and why.

First, ensure that there is electricity getting into the control circuit from the transformer or thermocouple. Whether the circuit is powered by 750 millivolts or 25 volts, the simple troubleshooting procedure is to follow the path of the electricity. Using your electrical multimeter testing unit, check to see if electricity is getting into each switch. If so, is it getting out? If not, then that is the faulty switch. As noted, you first determine whether the circuit electricity is powered at all.

On electronic ignition heaters, first check if there is current at the end of the circuit (Fig. 5-12). Leave the − (negative) lead touching the − of the transformer, and move only the + lead to test each part of the circuit. If there is current, then obviously all the switches in the circuit are closed, and you need to look elsewhere for failures. If there is no power, then one of the switches is open for some reason.

On standing pilot heaters, use the same tip. Check the beginning of the circuit at the combination gas valve to determine that the pilot generator is delivering 400 to 700 millivolts. Touch the − lead of your meter to the − terminal on the gas valve and the + lead to the + terminal.

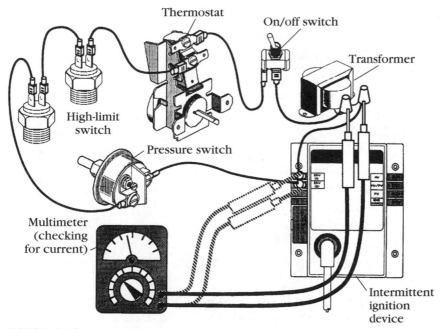

FIGURE 5-12 Testing control circuit. *Raypak, Inc.*

The heater will work on as little as 200 millivolts. But the electricity required to power the circuit and the combination gas valve will use almost all that, so the heater may fire, but it will soon shut down. A healthy pilot generator delivering over 400 millivolts will withstand the 200-millivolt operational drop and still retain enough power to continue the job.

Now check the end of the circuit. As noted above, leave the negative lead touching the negative (−) terminal on the gas valve, and move the positive (+) lead around the circuit. Again, if there is 200 millivolts or more there, the circuit is complete and the heater should fire. If not, start checking the control circuit switches.

Be aware that the switches themselves may be fine, but there may be a short circuit in the wiring. Sometimes rodents will nest in a heater and for some reason gnaw at the wiring. They rarely cut all the way through the wire, but by stripping the insulation, the bare wire sometimes comes in contact with the metal cabinet and creates a dead short—the electricity just flows through the cabinet and is dissipated or sent through the ground wire.

On a millivolt heater, the symptom of this is usually that even though the pilot generator is producing 400 to 700 millivolts, the pilot goes out when you release the gas valve control knob up after lighting. To test if this is the problem, remove both wires of the control circuit from the gas valve. If the pilot now stays lighted, you have a short in the wiring. If it still goes out, the gas valve is bad and should be replaced. On a 25-volt system, there is no simple way like that to test for bad wires, except visual inspection.

For millivolt or 25-volt systems, the control circuit switches are similar. They will also be the same for electric-fueled heaters, so use this troubleshooting section for those heaters as well. As noted above, follow the path of electricity and test for voltage into and out of each switch until you find the one that is open.

Another way to test a switch is to "jump" it. Take a short wire and connect the two terminals of a switch, in essence completing the circuit by bypassing the switch itself. If the heater fires, then obviously this switch is open.

As a time saver, you may want to test the last switch in the circuit first and work back to the first—when you find a switch with power to it, you know everything prior to that one is working, and it may save you checking several switches unnecessarily. Another time saver is to start with the most accessible switches first, since you may find the problem without "digging" into the cabinet (some high-limit switches and thermostats require face plate removal to test). You will develop your own preference, but I proceed with the switches most accessible (and most often the problem), as follows:

PRESSURE SWITCH FAILURE

RATING: ADVANCED

Most pressure switch failure is due to obstructions in the circulation—low water in the pool/spa; clogged skimmer basket or main drain; clogged pump strainer basket; or dirty filter. Check all these before you touch the pressure switch.

Sometimes the obstruction is in the water supply to the pressure switch. Remove the wires to the switch and unscrew the switch (Fig. 5-4, item 11) from this tube. Turn on the pump. Water should flow vigorously from the tube. If it does, look in the hole at the end of the pressure switch to be sure it is clear of obstruction.

If no water is coming out of the tube, remove the heater vent top, front header face plate, and flue collectors to expose the end of the tube in the header. Unscrew the tube and remove it. Blow through the tube to clear obstructions. Turn on the pump. Water should shoot out of the hole in the header if there is no obstruction.

If the problem is not in the water supply, adjusting the pressure switch to be more or less sensitive is done with the screw or knob on the switch. When you adjust either way, turn the screw only a quarter-turn at a time, as these switches are very sensitive. Check the operation after each quarter-turn until the heater operates correctly. The heater should fire and stay lighted when the circulation is running; it should shut down within 3 seconds when the circulation stops.

If the switch can no longer be adjusted or is rusted out, replace it.

FUSES AND FUSIBLE LINKS

RATING: ADVANCED

As previously described, some 25-volt heater designs include an in-line fuse, like a car fuse, in a bracket on the positive wire coming out of the transformer. Make a visual inspection or use your meter to test for current at a point after the fuse, to test it. If current is not passing through the link, it must be replaced; but remember, you want to determine why the fuse failed in the first place. Is it simply worn out, or is there abnormal heating in the area? After replacing it and restarting the heater, examine the burner components carefully. Look for the following:

- Is the venting adequate, or is there an overheating condition?

- Is there soot or debris on the burner tray, causing overheating somewhere?

- Is the gas pressure low, causing "lazy flame" (yellow flames that seem to "lick" out of the burner instead of strong, blue flames burning straight up)?

HIGH-LIMIT SWITCHES

RATING: ADVANCED

High-limit switches are slightly different in each make of heater, but the principles are the same. Remove the face plate or other protective cover (usually just one screw), and pull the switches with their retainer bracket from the header area. This operation can be done with the pump on or off. Pull the switches from the bracket, noting how this setup is assembled (pay attention, or it can be a jigsaw puzzle to put back together!). Pull the switch from its socket and replace.

ON/OFF SWITCH AND THERMOSTAT

RATING: ADVANCED

The mechanical version of thermostat switches can be tested for current just as the others can. The electronic versions look more complex, but are actually easier to diagnose because they are connected through a terminal block where all the connections are in one place for easy testing.

To avoid heat-related injuries, manufacturers limit their thermostats to 104°F (maximum temperature recommended for spas by most local codes and health laws). Since the mechanical thermostat has an accuracy rate of ±3°, these units are calibrated for 101°F (38°C) maximum to allow for an extra 3° inaccuracy. Of course that means it might be 3° the other way, and you may only get your spa to 98°F (37°C) or so. Now add bubbling water and a cool evening breeze, and the water only goes to the low 90s.

Electronic thermostats are calibrated to go no higher than 104°F (40°C), because their accuracy rate is within ¼ of 1°. Still, on a cool, windy night with the jets and blower going full-bore, the water may not get higher than the high 90s. Be aware that the problem may be in the design, not in the components or heater itself.

Replacing the mechanical thermostat requires removing the draft hood top of the heater to expose the oil-filled bulb and tube which are attached to the thermostat dial. Remove this along with the thermostat, and be sure the replacement is the same model, not one with a shorter tube that may not reach the dry well.

Replacing the electronic type means replacing the unit on the panel in the front of the heater only, unless you also suspect the heat sensor (thermistor). If so, the thermistor is found in the same location as the tube/bulb type in the header.

FIREMAN'S SWITCH

RATING: ADVANCED

A fireman's switch is simply an on/off switch attached to the time clock that shuts off the heater 20 minutes before the pump. The result is a cooldown of the heater before the water circulation stops, prolonging component life. You can troubleshoot this switch by jumping across the electrical wires leading to it at any convenient place. If it is defective, replacement is self-explanatory.

INTERMITTENT IGNITION DEVICE (IID)

RATING: ADVANCED

The IID is actually a sophisticated switching device, not an electronic "brain," as some call it. There are no hidden computer chips in this device. IIDs all have the same terminals, although perhaps not in the same location on the box. Figure 5-12 shows that power is supplied to the device by the transformer after passing through the control circuit. The IID sends power to the pilot to spark ignition and simultaneously opens the pilot gas valve by sending power to the appropriate terminal on the combination gas valve.

When the pilot lights, the heat creates a voltage which is detected by the IID and which acts as a signal to send power to the main gas valve on the combination gas valve, opening it to send gas to the burner tray.

The IID must be grounded to operate properly, so that is a good place to start your troubleshooting. If you have 25 volts at terminal 2 on the IID but no spark on the pilot, check the pilot. If it looks good, test the orange high-tension cord. The easiest way to do this is to remove the end of the cord from the IID and hold it about ⅛ inch (3 millimeters) away from its terminal on the IID. Turn on the heater and if you get a spark jumping from cord to terminal, the cord is probably good. If you get no spark, it is faulty.

If you get a spark here but not at the pilot, the IID is faulty. If you get a spark but no gas to the pilot, the problem may be in the high-tension cord (it may be sparking the pilot, but not sensing that the pilot has lighted). If replacing the cord doesn't solve the problem, the fault is in the IID.

Finally, if you have a spark and a lighted pilot but no voltage at the "MV" (main valve) terminal of the IID, the IID is faulty. I always carry a spare high-tension cord and IID. If I suspect either component, it is easy to replace it and locate the problem. If it turns out that the component is okay, then my spare just goes back in the toolbox and I put the original back in the heater.

IIDs, like combination gas valves, can be repaired at the factory but not by you. However, the cost of a new one ($70 to $100) is less than the labor, parts, and trouble of taking it to the factory for repair.

TRICKS OF THE TRADE: HEATER PREVENTIVE MAINTENANCE

RATING: EASY

- **The best preventive maintenance is to use the heater regularly. Corrosion, insects, rodent nesting, and wind-blown dirt create many heater problems that can be eliminated by regular use of the heater. The heat will help to dry up any airborne moisture which might otherwise start rusting the components. It will discourage insects and rodents before they get too "comfortable."**

- **Heaters need to be visually inspected from time to time. A look-around will detect sooting, gas or water leaks, or other problems before they begin.**

- **Keep leaves and debris off the top of the heater.**

- **Look at the pilot and burner flames. Are they strong, blue, and burning straight up at least 2 to 5 inches (5 to 13 centimeters)?**

- **Open a drain plug on the heat exchanger and look for scale build-up.**

Solar-Fueled Heaters

There are two ways to heat your spa at no cost after the initial investment. First, covers keep the heat in the water and can trap warmth from the sun during the day. Chapter 6 discusses covers in greater detail.

The second inexpensive way to heat your spa is to capture the power of the sun.

The concept of the solar panel is that it absorbs heat from the sun which is transferred to the liquid as it passes through, but the efficiency of a solar heating system is a factor of system setup. Exposure to direct sunlight, hours of sunlight, and amount of wind, clouds, or fog are all important factors in setting up the system which will impact its efficiency.

Solar panels are made from plastic or metal and are then glazed (covered in glass) or left unglazed. Obviously the glazed panel is heavier and more expensive; however, it absorbs and retains more solar heat and is therefore more efficient (fewer panels are required to accomplish the same amount of solar heating).

Figure 5-13 is called an *open-loop system,* meaning it is open to the spa water, mounted on the ground next to the pool or spa. Figure 5-14 also depicts an open-loop system mounted on a rooftop. Plumbing

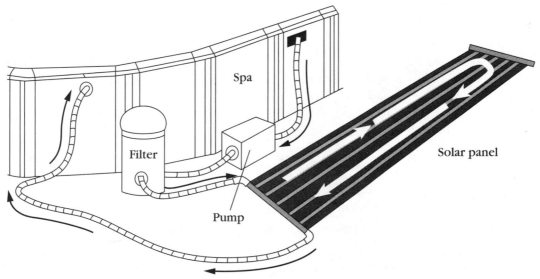

FIGURE 5-13 Solar heating diagram—on-ground.

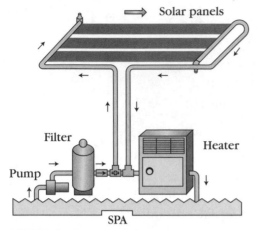

FIGURE 5-14 Solar heating diagram—rooftop. *Suntrek Industries.*

for solar heating is not different from other spa plumbing. It is located between the filter and the heater, so water going to the solar panels is free of debris and available for free solar heating before costly gas or electrical heating by the system's mechanical heater.

To Solar or Not to Solar?

To effectively heat with solar, regardless of the type of panel, the general rule of thumb is this: 75% of the surface area of the spa will be the surface area of panels needed. For example, if the spa is 50 square feet (4.65 square meters), you will need at least 35 square feet (3.26 square meters) of panels Since panels are generally made 4 feet by 8 feet (32 square feet, or 3 square meters) or 4 feet by 10 feet (40 square feet, or 3.72 square meters), our example spa would need one panel of either size to effectively heat the spa.

Next, a location must be found where the panels can face the sun. It may differ in your area, but generally the best will be facing south, to obtain the most hours of sun per year, winter and summer. Another factor is the prevailing wind in your area. High winds can tear panels from the roof or create so much cooling that the panels will not be effective. Such concerns will also dictate how many panels you need to install.

Finally, you will want to consider the cost. A simple solar panel system without automated controls will pay for itself in a short time, depending on how much you use your spa and, therefore, how much electricity or natural gas you are saving by using solar energy instead. That payback period will be significantly longer if you are adding automated controls, but those are usually reserved for large pool installations.

Installations

RATING: ADVANCED

If you decide to proceed, you might want to purchase a solar heating package from your pool and spa retailer that includes panels, plumbing, controls, and instructions. Simple flexible plastic panels (Fig. 5-15)

FIGURE 5-15 Flexible plastic solar panel.

are easy to add to your plumbing system and can lie on the ground, a deck, or a roof. This type of solar heating is very affordable, and each system will come with installation instructions for maximum performance.

If you choose a larger system, perhaps for a pool/spa combination, you may choose larger panels, which can be mounted on the ground (Fig. 5-16), or a more traditional rooftop installation (Fig. 5-17). In this case, you may want to hire a licensed carpenter (let her or him get the building permits and take the liability) to help with the installation and support of the panels, and you complete the plumbing into the equipment system.

Your spa's circulation pump will typically be adequate for moving water through the solar panels, unless they are located on a rooftop some distance from the spa. In that case you might need to upgrade the size of the pump or add a separate circulating pump just for the solar heating system.

Ball and check valves should be used so that the solar heating system can be completely and easily isolated from the circulation system, allowing normal spa operation when repairing the solar panels or when the desired temperature has been reached and no further circulation through the panels is needed.

FIGURE 5-16 On-ground solar installation. *Suntrek Industries.*

FIGURE 5-17 Rooftop solar installation. *Suntrek Industries.*

The angle of the panels (*pitch*) as they sit on the roof or ground is also important, because the more the sun's rays strike the panels at a 90-degree angle, the more heat will be absorbed into the water. Therefore, a 20- to 30-degree pitch helps the efficiency of the system in winter when the sun tends to be on the horizon rather than directly overhead. Aim the panels directly south or as close to south as is practical for the location of the panels.

As mentioned, there are numerous manufacturers and styles of panels and controls, far more than can be outlined here. Do some

QUICK START GUIDE: ADD SOLAR HEATING TO YOUR SPA

APPLIES TO ALL STYLES OF SOLAR PANELS

RATING: ADVANCED

1. Prep
   - Unpack solar heating kit: panels, plumbing, and connectors. Read the owner's manual.
   - Shut off pump and tape over switch or breaker, so no one can turn it on before you finish.
   - Isolate equipment plumbing by closing valves at suction line (at skimmer and/or main drain connection before pump) and return line (at spa discharge outlet).

2. Set Up
   - Mount panels per instructions of the owner's manual on ground, deck, prefabricated rack, or roof.
   - Aim the panel to the south for best performance.

3. Plumb
   - Cut spa equipment plumbing *after* the filter but *before* the heater (if your system has one).
   - Plumb solar panel "intake" line to discharge pipe from filter; plumb solar panel "outlet" line to spa return line (or "intake" line of heater if you have one). Use shut-off valves at each location.

4. Start Up
   - Reopen spa plumbing valves (and valves on solar panel plumbing).
   - Start pump, purge air, and check for leaks. The solar panels are now in operation.

research of what is available in your area. Just so you know what to look for, here are a few types:

- Plastic panels (glazed or unglazed)
- Metal panels (glazed or unglazed)
- Thin, lightweight aluminum panels
- Rubber panels (like doormats) that nail directly to a rooftop
- Flexible plastic or metal hose, coiled on a rooftop or built into a concrete deck (in this case, the sun heating the deck in turn heats the solar coils)

Whatever the style, remember when you plan an installation that the pipes to and from the panels should be insulated so that heat is not lost along the way.

Maintenance and Repair

RATING: EASY

Once the system is installed, most homeowners and pool technicians tend to forget about the solar heating system. Inspection every 2 or 3 months should be made, however, to check for leaks. Leaks can easily occur because of the extremes of hot and cold temperatures which cause the panel materials to expand and contract. Leak repair will depend on the type of material in the panel or plumbing, and each manufacturer makes leak repair kits with instructions. Obviously, the plumbing to and from the panels can be repaired as needed by using techniques outlined in Chap. 2.

The second common problem is dirty panels. Dirt prevents the panels from absorbing heat and can cut efficiency by as much as 50%. Regular solar panel cleaning, requiring no more than soap and water, is the best way to ensure efficient operation and a hot spa.

Now that we have covered the most common problems, here is a troubleshooting guide showing symptoms and solutions that may assist in quickly identifying and repairing solar heating difficulties:

Spa is not as warm as it should be.

- Panels are too small or incorrectly oriented.

- Circulation through panels is not long enough each day.
- Circulation is at wrong time of day (if water goes through the panels at night, the water may be cooling instead of heating).
- Panels are dirty.

Air bubbles are found at spa return lines only when solar is operating.

- Check for clean filter.
- Vacuum relief valve is not operating properly or is clogged.

Some panels are warm to the touch, others cool.

- Check for circulation problems.
- Check for any valves between panels.

There are leaks in panels or plumbing.

- Check water chemistry.
- Check that panels and plumbing are secure.

By the way, don't be concerned that plastic panels are not functioning properly because they are cool to the touch. When water is circulating through the panel, it will transfer the heat to the water and away from the plastic surfaces. If the solar panel has been shut off for some time, the panel will be hot to the touch until water begins to circulate. Be careful to warn bathers to stay away from the return water discharge port in the spa when reopening the solar heating system, because the water that has been sitting in the panel can be scalding.

Automation and Optional Equipment

After you have circulated, filtered, and heated the spa water, you may want to add therapeutic bubbles, lighting, or a cover. You may also want to have some of your spa's features operate automatically or by remote control. This chapter will take you through the details of these additions to the modern spa or hot tub.

Time Clocks

Time clocks ensure that water is circulated through the filter each day and that the spa is hot at the time you intend to use it. They can also be used to turn lights on and off or operate other equipment automatically. This section deals only with mechanical timing devices, since electronic timers are typically part of portable spa control units which also include electronic thermostats and other sophisticated controls. Electronic controls of all types will therefore be described in the section "Spa Automation and Remote Controls."

Electromechanical Timers

Figure 6-1 shows of a typical 120-volt time clock. This time clock allows various "on" or "off" settings throughout each 24-hour period. Some models include a 7-day feature, allowing you to determine which days of the week you want the clock to control the system.

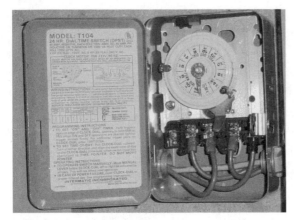

FIGURE 6-1 Electromechanical time clock.

FIGURE 6-2 Time-setting tripper or "dog."

Figure 6-2 shows a closeup of the on and off trippers (the word *on* or *off* is etched on the tripper). When you set these over the desired times on the face of the dial, the clock will control the flow of electricity to your system accordingly. More than one set of trippers (also called *dogs*) can be affixed to the dial, so the system could go on and off several times during a 24-hour period.

To set the clock, you simply pull the face toward you and rotate it until the number on the dial, corresponding to the correct time of day, is under the time pointer. By pulling it forward, its drive gear is disengaged from the motor drive gear and will freely rotate. The time pointer is the only "hand" on the clock. The dial is divided into each of the 24 hours of the day (differentiating between "a.m." and "p.m."; some clocks say *Day* and *Night*), and each hour is further divided into ¼-hour segments, so when setting the correct time of day, you can be fairly precise. Thus, the time clock needs only one hand or pointer to tell you the time.

After you locate the correct time of day, release the dial, and it will snap back into position, reengaging its drive gear with the motor drive gear, to operate.

The switch lever (seen just below the "on" tripper in Fig. 6-2) moves to the right, and it lifts the contacts apart, breaking the electric circuit. When the lever moves back to the left, the contact arm is allowed to drop back into contact with the lower contact, completing the electric circuit. The lever also allows manual operation, so regardless of where the trippers are set, you can manually operate the system.

The contacts are attached to screw terminals for attaching wires of the appliance to be controlled. On this 120-volt example, the hot 120-volt wire is attached to the terminal marked *line* and the neutral to its terminal. The hot wire for the load (the appliance) is connected to the *load* terminal, and its neutral also is connected to the *neutral* terminal. Some clocks provide separate neutral terminals for the line and loads; however, they both join to the same wire that returns to the circuit breaker. The 240-volt version includes two line terminals for the two hot lines coming in and two load terminals for the supply to the appliance.

Note that the wires feeding electricity to the clock are wrapped on the line and neutral terminals (with a 240-volt clock, the one lead would be attached to each line terminal). This way, there is a constant supply of electricity to the clock.

A quick way to tell if power is getting to the clock (assuming the clock motor is in working condition) is to look through the opening marked "Visual Motor Check" (Fig. 6-3). Look in here to see the motor gears

FIGURE 6-3 Time clock motor inspection hole.

spinning, to verify that something is operating! Some clocks have the motor mounted on the front of the clock, and the center hub of the drive gear is visible. Close inspection here will reveal whether the hub is spinning.

Waterproof boxes in metal or plastic are built to house time clocks. The unit shown in Fig. 6-1 simply snaps into clips built into the box. Other versions have built-in brackets that are designed to align with screw holes on the clock plate, and these are held in place by two machine screws. The boxes are built with knockout holes to accommodate wiring conduit of various sizes and predrilled with holes in the back for mounting the box to a wall.

FIGURE 6-4 Twist timer.

Twist Timers

Twist timers are used mainly for booster motors, lights, or air blowers when only a limited amount of operation is needed. Twist timers are built to fit in a typical light switch box and contain no field-serviceable parts. Figure 6-4 shows a unit calibrated for 1 to 60 minutes, but others are calibrated for 1 to 15 minutes, 1 to 12 hours, and several variations in between.

The knob attached to the shaft that comes through the faceplate is twisted until its pointer or arrow aligns with the desired number of minutes. The circuit is completed in any position except off, and the mechanical timer is spring-loaded to unwind for the number of minutes or hours selected. When the spring is unwound, the circuit is broken and the appliance shuts off. Twist timers are available in 120 or 240 volts and are used in place of a simple on/off switch, but where users might forget to turn off such a switch.

Since twist timers are inexpensive and there are no serviceable parts, if one fails, replace it. It is no more difficult than replacing a light switch, and all you need is a screwdriver. When you buy the replacement, follow the instructions in the box if you don't find it self-explanatory, and as with any repair of an electrical item, be sure the electricity is disconnected at the breaker panel.

Repairs

There's not much that goes wrong with electromechanical time clocks, and it's easy to service most of them.

REPLACEMENT

RATING: EASY

When a time clock rusts out or the gears wear out and the unit needs to be replaced, it takes longer to buy the new unit than to perform the replacement.

When you buy the replacement, be sure to get the same-voltage clock as you have in the existing installation. Also, buy the same make of clock, since a different manufacturer's clock probably won't fit in the existing box. Finally, be sure to buy the same model—remember that some electromechanical clocks are built for 24-hour calibration, others for an entire week, and twist timers come in a variety of time calibrations.

1. **Power** Shut off the power supply at the breaker, and tape the breaker so no one inadvertently turns it back on.

2. **Disconnect** Remove the line and load wires (Fig. 6-1) from the old clock and any ground wire. Remember (or mark) which wire is which.

3. **Replace** Unscrew the holding screws or simply unclip the old clock to remove it from the box. Snap or screw the new one in place.

4. **Connect** Reconnect the new wires to the appropriate line or load terminals. Terminals on the new clock may not be

> **TOOLS OF THE TRADE: SPA CONTROL REPAIR**
>
> - **Flat-blade screwdriver**
> - **Phillips screwdriver**
> - **Needle-nose pliers**
> - **Nut drivers**
> - **Multimeter**
> - **Thermometer**
> - **Electrical tape**

arranged the same as in the existing unit, so be sure you are getting the *line* wires (two hots or one hot and one neutral) attached to the *line* terminals and the appliance wires attached to the *load*. If it is not clearly marked, the simple way to tell which terminal is which is to notice where the clock motor is connected—the two clock wires are always attached to the *line* terminals.

5. **Test** Turn the power back on, and test the clock operation by turning the manual on/off lever to on.

Keep in mind that the clock motor itself must always have a source of power, even though the time clock is turning the equipment on and off. If the time clock power supply comes from an on/off wall switch or other device, the power might be interrupted, meaning the clock cannot keep accurate time. If your installation has this type of intermediate switch, tape over it to keep it from being shut off, or remove it from the system entirely.

CLEANING

RATING: EASY

Sometimes insects will nest in a time clock. The cure is simple—turn off the power to remove and clean the contacts. When you have finished, apply a liberal amount of insecticide to the interior of the housing box before reinstalling the mechanism. Don't spray the mechanism itself. You may create unintended electrical contact through the liquid.

Other than that, corrosion will occasionally bind up a clock. If you know it is getting power but cannot see the gears turning, take the tip of your screwdriver and gently force the gear (the one visible through the inspection hole in Fig. 6-3) in a clockwise direction. That will often get it started, but give the gears a good general lubrication to prevent recurrence of this failure.

To lubricate a time clock, shut off the power source, remove the clock from the box with the wires still attached, and expose the rear of the clock. Apply a spray lubricant, such as WD-40, liberally around the gears. Do the same on the front side to lube the gears behind the dial face. Be careful not to wet the electrical contacts. Put the clock back in the box and turn the power back on. Turn the clock on and off a few times to work in the oil. If the clock fails again, replace it.

MECHANICAL FAILURES

RATING: EASY

The most common mechanical failure is caused by improper gear engagement when you set the time on the clock. When you pull out the dial face of the clock to set the time of day, take care as you release it that you are getting a true reengagement of the dial with the motor drive gear. Try setting different times on the clock, and you will note that sometimes the two gears don't mesh, but rather the dial gear sits on top of the motor gear. Obviously, the clock won't work like that.

The answer is to slightly wiggle the dial face as you release it. Don't just pull, turn to the desired time of day, and then release. As you release, twist the dial back and forth very slightly in your hand, to make sure the gears mesh. With a few practice settings you'll feel the difference between a dial that has gone back into place completely and one that is slightly hung up. This problem also occurs if you are trying to set the time of day close to a tripper that has already been secured to the dial. The tripper obstructs clean engagement of the gears, so it's best to remove the trippers when setting the time of day, then replace them after the gears are fully reengaged.

The second mechanical problem of time clocks lies with the trippers themselves. If the screw is not twisted tightly on the face of the clock, the trippers come loose and will simply rotate around the dial, pushed by the control lever instead of doing the pushing themselves. Check the trippers regularly because they can come loose over time or simply from system vibration.

Finally, even something as simple as a tripper can wear out, so if the clock is keeping good time but not turning the system on or off, try installing new trippers.

SETTINGS

RATING: EASY

When setting on/off trippers, you can apply them on the dial face side by side to what appears to be only 30 minutes between them. I have found that when they are too close like that, they won't operate the lever. Generally, trippers must be at least 1 hour apart to operate.

Make frequent checks of your time clocks. City power outages, someone working on the house who shuts off all the breakers for a time, the twice yearly daylight savings time changes, and any number of other household situations can interrupt power to the time clock. Each time that happens, the clock stops and needs to be reset when the power returns, in order to reflect the correct time of day.

Spa Automation and Remote Controls

Modern technology can make many spa and hot tub operations and maintenance tasks easier or completely automated, including filtration, heating, and operation of jets, lights, or bubbles. There are many types and manufacturers of these controls, but if you master the typical system described in this section, you will probably be able to operate and troubleshoot your own unit. It is also important to note that failures of various spa components are often related to electronic timers or other automated controls that are simply not set correctly.

The advantages of electronic controls are precision and low voltage. Some units are designed with such low voltage that you are permitted to install them next to a spa. Most have digital readouts so you can precisely set the time and temperatures desired as well as program a host of other features.

Remote control devices allow you to operate pumps, heaters, lights, blowers, and other devices without actually going to the spa. Such devices are also available with switches at or in the spa, so you can control appliances without getting out.

Remote control devices fall into two categories: those operated by pneumatic (air) switches and those operated by electronic switches (wireless or hardwired). We will examine remote controls first, because some of their basic operating principles are applicable to the more complex world of automated controls.

Air Switches

You have probably driven into a gas station, over the black rubber hose, and heard the bell in the station. Spa air switches operate on the same idea—by compressing air in a hose, you force a switch to go on or off.

Figure 6-5A shows a typical air switch. By depressing the plunger in the middle of the button, air is forced into a flexible plastic tube

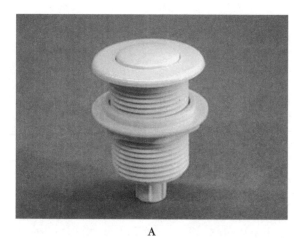

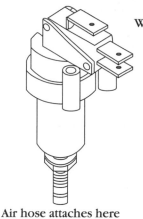

Wiring to appliance
attaches here

Air hose attaches here

A   B

FIGURE 6-5 (A) Spa air switch button. (B) Air switch components.

(usually ¼-inch diameter or less). The force is transmitted along the air tube until it reaches a simple electromechanical switch (Fig. 6-5B) at the other end, turning it on, off, or to another position. In this way, you can mount the button near or even in the spa and not worry about electricity near water. Air switches operate up to 200 feet (61 meters) from the button, and the air tube is protected by running it through a flexible conduit, similar to that used for electrical wires in your walls.

The most basic of these are simple on/off switches. Some activate a rotating, ratcheted wheel that in turn activates up to four switches. Thus, you push the button once to turn on the circulation pump and heater, a second push to add a blower, a third push to add a booster, and a fourth push to turn it all off. Most of these "four-function" units allow combinations that vary the pattern, depending on your use preference or what additional equipment you have.

The electrical part of the air switch is usually located on a waterproof box that houses the wiring connections, the actual switches (usually microswitches), and the terminals for attaching line and loads just as in a time clock. In some units there is a small electromechanical time clock that can operate one or more appliances on preset times in addition to the activation by the air switch button.

Little can go wrong with air switch units, and they require virtually no service. Here are a few pointers, however.

TROUBLESHOOTING

RATING: EASY

When an air switch fails, you need to determine if the problem is with the button, the hose, or the electrical switch. To isolate the problem, I carry a piece of hose with me (about 3 feet or 1 meter long). I disconnect the air switch system hose from the switch nipple (inside the electrical component box of the switch system) and place my hose on the nipple and blow through it. If the switch operates normally, I know the problem is not there. If it doesn't, I know the problem is not likely in the hose or button. Similarly, if I suspect a defective button, I can disconnect it and attach it to my test hose. If it activates the switch, I know that both the button and switch are okay, so the problem must be in the existing hose.

Rarely, the microswitch itself will wear out; it can be bought separately and replaced (an operation that is self-explanatory after you remove the faceplate of the switch box to reveal the actual microswitch location—usually only two screws and a couple of bayonet wire terminals are involved, much as in replacing a time clock, as described above). Don't forget that if your air switch system includes a time clock, the failure might be related to the on/off setting of the time clock and not to the air switch components themselves.

AIR LEAKS

RATING: EASY

Using the troubleshooting method just described, what can you do if you determine that the failure is in the air hose between the air switch button and the electrical switch box? If you can reach both ends of the hose, just tape the new hose to the old (starting at either end), then pull the old hose out. When you get to the taped joint, you have just snaked a new hose into the conduit. If you cannot access one of the hose ends, follow the replacement procedure detailed below.

REPLACEMENT

RATING: EASY

Sometimes the air buttons wear out, but these are easily replaced. Some are designed with a *collar* that is mounted into a deck or spa wall and a removable button. Buttons are fitted with a large nut on the underside (Fig. 6-5A) so you can mount them on a deck and tighten up the nuts to hold the collar in place.

When you remove the center button portion, if there is no slack in the air hose, it will come off the end of the button and may be hard to reach for attachment to the new button. Try pushing the hose from the other end (the end attached to the electrical switch box) toward the button to force it back out of the conduit. If there just isn't enough hose, pull out the old hose and run an electrician's fishtape through the conduit. When the end comes out, tape the new hose to it and pull the fishtape back through. Now you can attach the hose ends to the button and the switch.

NEW INSTALLATION

RATING: ADVANCED

Original installation of an air switch system is not difficult, but you may want to hire an electrician to help connect the wiring to your breaker panel. To save time and money, you may want to buy the system, mount the electrical box, run the hose in conduit, and mount the button. Then call in the electrician to do the electrical connections.

You will find that not all air hoses are run in conduit. Because they have no air or electricity in them, there are no standards or rules about running a length of hose. When you are making a new installation, I strongly recommend running the hose through a ½-inch (13-millimeter) PVC electrical conduit. This will protect the hose from the elements and keep it supple. It will also keep out rodents or household dogs that might want to chew the hose.

Use electrical PVC conduit and fittings rather than plumbing PVC. If you have to use a fishtape to snake in a new line, the tape will get hung up on the sharp angles of the plumbing elbows and connections, whereas electrical connections are "swept" gradually into the elbow or angle to make that less likely.

Wireless and Hardwired Remote Controls

Wireless remotes are composed of a sending unit (Fig. 6-6, center) with anything from 4 to 24 buttons in a waterproof, battery-powered case (low voltage for safe waterside use). The buttons send a signal to a receiver (Fig. 6-6, left) that might be as much as 1500 feet (457 meters) away. That allows the sending unit to be in the house, for example, while the

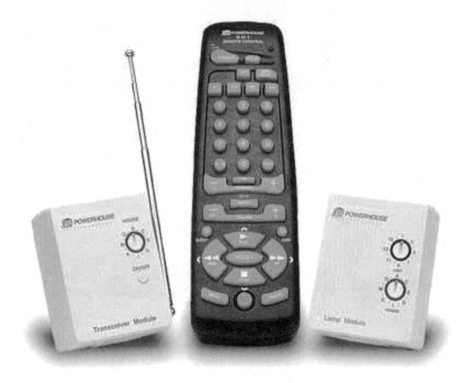

FIGURE 6-6 Wireless remote control and receiver.

spa might be some distance away in the yard. Another application of this technology is to use the sending unit at the spa to control equipment that is located some distance away.

The receiver is either hardwired to the electrical supply or plugged into a standard household electrical outlet. The receiver "interprets" which button has been pressed and activates a switch which turns that piece of equipment on or off. In essence, it works just as the air switch does, except the signal is sent by radio signal instead of compressed air. The pump, heater, air blower, lights, or other appliances are plugged into the receiver or hardwired to it.

Some button units do not have batteries, but plug into any household outlet (Fig. 6-6, right). When you press the button, the signal is sent along the household wiring to the receiver, wherever it is located, which is powered by the same household current. Obviously, since these sending units are powered by 120-volt household current, they cannot be located near the body of water. The value of these is that they are not subject to weak-battery failure or weak radio signals, which sometimes fail to penetrate thick walls or long distances.

The 120-volt remote control is usually installed where the customer plugs the sending unit into an electrical outlet in the house to be able to turn on and heat up a spa that may be located out in the yard. The disadvantage, obviously, is that to turn appliances on or off while using them, you must get out of the spa.

TRICKS OF THE TRADE: GENERAL TROUBLESHOOTING GUIDE FOR SPAS (SYMPTOMS AND SOLUTIONS)

The following will help in diagnosing general equipment failure before you look at more challenging control system failures. Always try the simple things first!

Spa won't run (all controls failed).

- Check circuit breaker and reset button on the ground fault circuit interrupter.
- Check equipment access door—it may have circuit breaker when door is ajar.
- On portable spas, check that outlet is hot and that plug is properly inserted.

Spa runs, but jets are weak.

- Clean the filter.
- Water level is too low for skimmer or other suction line. Refill.
- Suction lines or pump is clogged with hair or other debris. Inspect and clean.
- Air may be trapped in the system. Bleed air from filter or other points in system.

Water is not getting hot.

- Check thermostat setting.
- Check reset button on heater.
- Check circulation. If it is weak, pressure switch may have turned off heater.
- Check length of heat/circulation time.
- Make sure system is in heating mode (some systems have a filtration mode without heat).

There are no bubbles.

- Open air vents on jet plumbing to introduce air into return lines.
- Check that blower is plugged into the control panel.

Wireless remote systems are very simple to install. As mentioned, the receiving units can be plugged into a household electrical outlet, and the appliance plugged into the receiver. The appliance can now be operated from the remote control—it's really that simple. Other models use hardwiring rather than plug-ins, but the concept is the same. Hardwired receivers are installed wherever the spa equipment is located, simply intercepting the wires that feed the appliance and running them through the receiver unit.

The main failure of these units occurs when the battery becomes weak, and replacing the battery solves the problem. The other common failure is that the radio signal cannot penetrate thick walls or long distances between the spa and equipment location. In this case, you may need to use the remote sending unit that plugs into the household electrical outlet.

The same remote control components can also be hardwired from sender to receiver to appliance. In this case, you are well served to hire a professional electrician to install or troubleshoot your system. A hardwired remote can also be as simple as a standard on/off wall switch in the home that turns the spa pump and heater on or off. When more control features are needed, or when more pieces of spa equipment are being controlled, most spas today will employ sophisticated low-voltage automated units which control relay switches to power each function.

Automated Controls

Each make and model of automated controls will differ in setup, nomenclature, features, appearance, and repair. But all systems have basic elements in common, which means that if you can install, operate, troubleshoot, and repair one, you will probably be successful with any other system. The common features of automated control systems include the following:

Automated controls typically command

- Filter pump
- Jet pump
- Air blower
- Lights

- Heater (and temperature)
- Other auxiliary equipment

Standard 110-volt or 220-volt power feeds relays that actually energize each piece of equipment.

Relays are controlled by low-voltage (less than 24-volt) switches that are remotely activated or hardwired.

Low-voltage switches are operated manually or by sophisticated computer controls.

Display panels allow for programming of function, timers, temperatures, and troubleshooting.

Today's automated controls are more complex than the ones of just a few years ago, but the good news is that they offer many more features and are organized like familiar computer programs. Widespread computer literacy has also led the spa industry to create owner's manuals that are more user-friendly, including easy-to-follow "quick start" menus and more "plug and play" control components that can be easily added or simply unplugged from the system and replaced when they fail.

Figure 6-7 shows the power center (with the safety panel removed to show the relays and wiring) of a popular automated control system, the Aqualink, made by Jandy. The control panel, also called the printed-circuit board (item 1); relays (item 2); and 110/220-volt circuit breakers (item 3) are housed in a weatherproof metal box (item 4) with the wiring diagram (item 5) pasted inside the door. The power center is programmed to operate the spa equipment automatically or manually and is therefore the heart of any automated system.

The power center's control panel is the true brain of the system, constantly scanning the system to determine if an action is required. For example, it senses a change in water or air temperature and turns the heater on or off. It senses that a manual button has been depressed at the spa and activates the specified piece of equipment.

Figure 6-8A shows the same power center with the control panel removed. Now you can see how the electricity moves from the breakers through the relays to the spa appliances. A relay is simply a switch which allows household current (110- or 220-volt) to power an appliance. The relay switch closes, completing the circuit, when it is energized

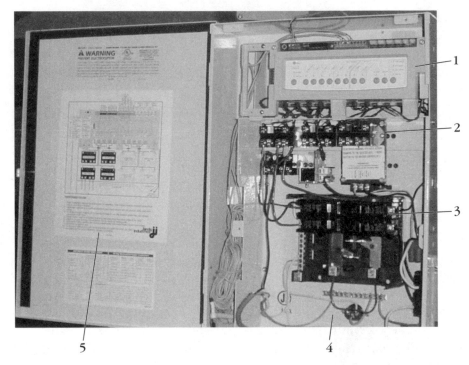

1 Control panel
2 Relays
3 110/220-volt circuit breakers

4 Weatherproof box
5 Wiring diagram

FIGURE 6-7 Automated control power center.

by a low-voltage magnetic coil. In this way, safer low-voltage controls can be used to turn high-voltage appliances on or off.

Figure 6-8B shows the functions of the control panel and the wiring diagram that is pasted inside the power center door. While the control panel can be programmed to automate all spa functions, the power center is usually located in an equipment area some distance from the spa. The spa itself may be located some distance from the house. Therefore, remote controls are added to the system that allow manual operation or reprogramming of automated functions from more convenient locations, either in the home or from spa side. These remote controls can be hardwired or wireless.

Figure 6-9 shows a hardwired spa-side remote control, an in-home remote control, and a wireless remote control. All these are designed to work with the power center shown in Fig. 6-7. The hardwired units are designed as modules, typically four pin connectors, that plug into the power center and require no special connection, soldering, or other wiring. The spa-side unit is water-resistant, although not designed to be mounted underwater. There are also simple push-button models that can be mounted spa-side. Remotes are powered by 10 volts of direct current (DC).

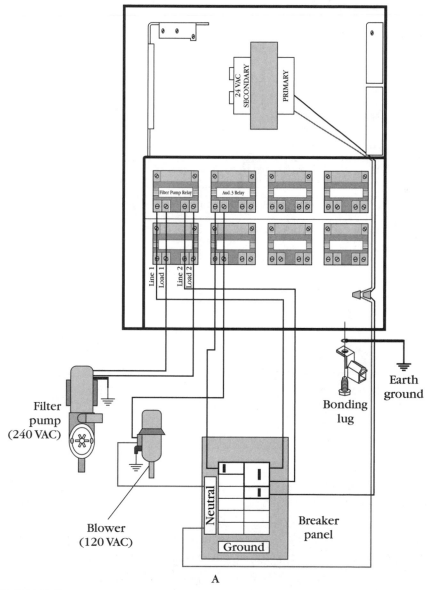

FIGURE 6-8 (A) Wiring of power center. (B) Close-up of power center control panel and wiring diagram. *Images courtesy of Jandy.*

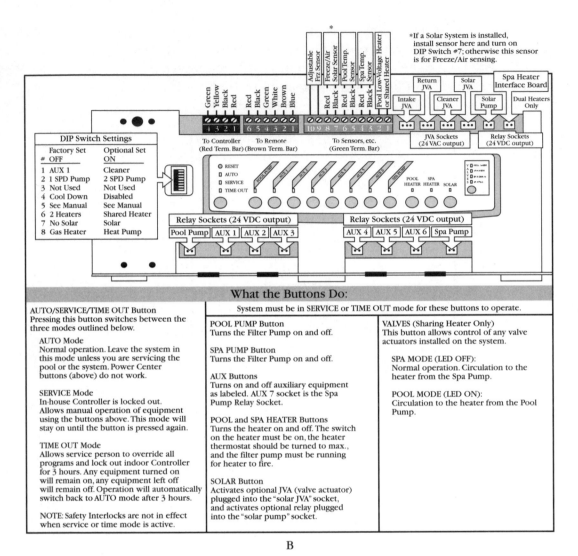

What the Buttons Do:

System must be in SERVICE or TIME OUT mode for these buttons to operate.

AUTO/SERVICE/TIME OUT Button
Pressing this button switches between the three modes outlined below.

AUTO Mode
Normal operation. Leave the system in this mode unless you are servicing the pool or the system. Power Center buttons (above) do not work.

SERVICE Mode
In-house Controller is locked out. Allows manual operation of equipment using the buttons above. This mode will stay on until the button is pressed again.

TIME OUT Mode
Allows service person to override all programs and lock out indoor Controller for 3 hours. Any equipment turned on will remain on, any equipment left off will remain off. Operation will automatically switch back to AUTO mode after 3 hours.

NOTE: Safety Interlocks are not in effect when service or time mode is active.

POOL PUMP Button
Turns the Filter Pump on and off.

SPA PUMP Button
Turns the Filter Pump on and off.

AUX Buttons
Turns on and off auxiliary equipment as labeled. AUX 7 socket is the Spa Pump Relay Socket.

POOL and SPA HEATER Buttons
Turns the heater on and off. The switch on the heater must be on, the heater thermostat should be turned to max., and the filter pump must be running for heater to fire.

SOLAR Button
Activates optional JVA (valve actuator) plugged into the "solar JVA" socket, and activates optional relay plugged into the "solar pump" socket.

VALVES (Sharing Heater Only)
This button allows control of any valve actuators installed on the system.

SPA MODE (LED OFF):
Normal operation. Circulation to the heater from the Spa Pump.

POOL MODE (LED ON):
Circulation to the heater from the Pool Pump.

B

FIGURE 6-8 (*Continued*)

Like many automated systems, the one shown in Fig. 6-7 is designed with several important features:

■ The programming is done from the remote control panel in the home. A service module is sold separately, which is the same control unit but one that can be plugged into the power center directly when you are performing service or repairs. When plugged into the power center, the service module can lock out control from other remote locations to ensure that only one set of commands is being given to the programmable circuit board.

A

■ The control panel in the power center contains the memory of all programming and is powered by the incoming electricity from the breakers. This unit also employs a backup 9-volt battery to prevent memory loss in case of power outages.

■ The power center is also served by air and water temperature sensors, called *thermistors,* which measure temperature and send the information via hardwiring back to the control panel.

B

■ Control panel and remote control buttons, along with the relays they control, are typically labeled *Filter, Heat,* and several *Auxiliary* buttons. Auxiliary buttons are connected to other optional equipment, such as an air blower, jet pump, spa lights, outdoor lights, or pool equipment (when the spa is part of a pool/spa combination installation).

■ Spa-side remote controls include a small heating element to drive out moisture, but the units themselves are not meant to be mounted below the waterline of the spa.

■ Automated control systems are also designed to operate motorized valves when a spa shares equipment with a pool. For example, the system can

C

FIGURE 6-9 (A) Spa-side remote control. (B) In-home remote control. (C) Wireless remote control.

TRICKS OF THE TRADE: TROUBLESHOOTING AUTOMATED CONTROLS

RATING: ADVANCED

Each make and model of automated control comes with an owner's manual that provides flowcharts and decision trees to make troubleshooting easy. Start by consulting these charts. There are a few problems that are common to most makes and models of automated controls.

Remote control unit fails to operate the selected appliance.

- Check circuit breakers and manual switches at power center to ensure all are on.

- Check programming to ensure that no timer or other overrides are controlling the supply of power to the appliance.

- Attempt to operate appliance from the power center control panel. If appliance operates from power center, the fault is in the remote control unit or wiring between the power center and remote control unit. Note that most hardwired remote control units do not operate at more than 300 feet (90 meters) from the power center because of voltage loss along longer wire runs (especially on cold days).

Heater is not operating or fails to maintain desired temperature.

- Thermistor failure is common to automated systems. Thermistors are sensors that are placed in the water stream, usually in the plumbing after the heater, and send signals back to the control panel in the power center. Your owner's manual will show you where and how to test these sensors, using a multimeter electrical tester (Fig. 6-10).

- The manual thermostat on the heater itself must be set at the highest temperature setting possible for the remote control to operate properly. Otherwise, this mechanical switch will shut the heater off when its setting is reached, rather than waiting for the automated control temperature setting to be achieved.

There are unusual or no displays on remote control units.

- Most automated controls have a Reset button in the power center, and many have in-line fuses in the remote controls themselves. Try pressing the Reset button, and if that does not reboot the system, look for a blown in-line fuse.

- Automated controls also typically have a 9-volt backup battery to maintain programming during power failures. Malfunctioning spa equipment can sometimes be traced to backup batteries that simply need to be replaced.

  In general, a thorough reading of the owner's manual will solve any problem you are likely to have with your automated controls. If the manual is lost, most manufacturers now provide them on their websites for easy reference and downloading. Some manufacturers also offer online troubleshooting assistance.

switch the valves from pool filtration and heating to spa filtration and heating and can change the thermostat setting of the heater from the desired pool temperature to the desired spa temperature automatically.

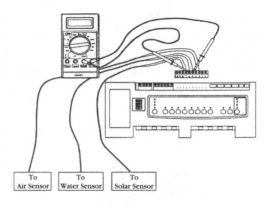

To Air Sensor | To Water Sensor | To Solar Sensor

Remote controls increasingly use computer-style programming. The remote control shown in Fig. 6-9A has a button for each function, while the unit in Fig. 6-9B and the wireless unit in Fig. 6-9C have a small screen and programming keys. There are also in-home remote control panels, called *one-touch* units, that have computer-style menus and programming. These are easy to follow and program, thanks to on-screen prompts and self-explanatory menu options.

If you are installing a new spa or adding an automated system to your existing spa, you will follow the installation instructions that come with each make and model. Typically, installation requires basic carpentry skills to mount the remote control units in the home or at spa side. You may want to have a professional electrician assist with the wiring and mounting of the power center, but other than that, installation of these units is quite simple.

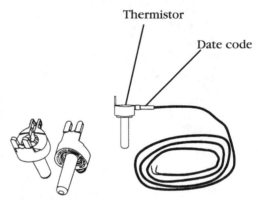

Thermistor

Date code

FIGURE 6-10 **Testing remote sensors.** *Image courtesy of Jandy.*

Ground Fault Circuit Interrupter

When equipment or wiring fails, it may draw more current than the appliance can use, burning out the appliance. The circuit breaker is designed to break the circuit when demand exceeds the rating of the breaker. It takes so little current to kill a human that the typical breaker would deliver lethal doses before breaking the circuit. In other words, circuit breakers are designed to protect equipment, not humans.

The ground fault circuit interrupter (GFI), on the other hand, is designed to protect humans. It is the circuit breaker that will detect problems at a low enough level to protect you before lethal doses are delivered. The GFI constantly measures the current going out of it, to

A

B

FIGURE 6-11 Ground fault circuit interrupter (A) outlet, (B) plug.

the appliance, and coming back into it. If an inadvertent grounding takes place, such as if the metal case of the appliance were electrified and you touched it, completing a pathway for current to the ground, the GFI will detect the drop in current it is receiving and will break the circuit. A GFI will detect variations as low as 0.005 amp (5/1000 of 1 amp), which is about one-half the lethal charge to a child and about one-sixth a lethal dose to you. The GFI cuts the circuit within ¹⁄₄₀ of 1 second, so it is not only sensitive, but also quick.

There are three basic styles of GFIs that you will likely encounter in spa installations. The first looks like a standard circuit breaker in the electrical panel, but it has a test button in the face of the breaker in addition to the on/off breaker switch itself. By pressing the test button, you are simulating the unbalanced-current condition inside the breaker and thereby testing the efficiency of the GFI. The GFI breaker resets the same way a normal panel breaker does.

The second type of GFI is built into a wall outlet, such as the type you might install for plugging in a portable spa (Fig. 6-11A). It will also contain a test button and switch to reset the GFI. Finally, you can purchase a "portable" GFI, a unit which plugs into a wall outlet; then the appliance is plugged into the GFI, making that outlet a GFI outlet. This variety is also incorporated into the plug of some portable spas (Fig. 6-11B). All types of GFIs, like any other mechanical device, are subject to failure and should be tested at least every month. Slight variations may occur in pool or spa equipment and cause the GFI to trip off even though everything is properly functioning. Panel GFI breakers are somewhat prone to this problem, since there is often slight current "leakage" inside the panel and long wire runs to the equipment location can create slight variations in current that will cause the GFI to trip. For this reason, it is best to locate the GFI as close to the appliance(s) as possible.

If a GFI is breaking the circuit, you troubleshoot the problem in the same manner as for any other breaker. Start by disconnecting the appliance and resetting the breaker. If it pops off, the problem is in the GFI. If not, the problem is in the appliance.

Electrical outlets located within 15 feet (5 meters) of the water's edge must be protected by a GFI, and circuits for all underwater lighting must be so equipped. All portable spas should be protected by a GFI.

Lighting

Spa lighting can be provided by simple, old-fashioned devices and more modern high-technology units. A basic understanding of each will allow you to light your spa for any mood or utility.

Standard 120/240-Volt Lighting Fixtures

Figure 6-12 shows a typical spa light fitted into the wall of a portable spa. Larger versions for gunite spas are housed in a stainless steel cone-shaped fixture, in a variety of sizes up to 8 inches (20 centimeters) in diameter by 6 to 10 inches (15 to 20 centimeters) deep. In both cases, the fixture itself is mounted in the wall of the spa in a container called a *niche.*

Like the lights in your house, spa light fixtures have a standard, screw-in socket for a bulb. The cord that supplies electricity to the fixture is waterproof and enters the unit through a waterproof seal. There are no user-serviceable parts in the cord, seal, or fixture except for the bulb. Fixtures and bulbs are available in 120 or 240 volts, but 120 volts is most common for residential use. Bulbs generally run from 300 to 500 watts.

Fixtures for portable spas are sold with a short wiring harness to reach the electrical

FIGURE 6-12 Typical spa lighting.

junction box within the cabinet. Fixtures for larger spas, where the junction box may be some distance away in the yard, are sold with cords of 10 to 100 feet (3 to 30 meters), so you need to know the distance from the light niche to the junction box to determine which is best for you when purchasing a replacement fixture.

Many smaller light fixtures for spas employ smaller-base specialty bulbs that may screw in or have "bayonet" type of sockets. To get the most light out of the lowest wattage and smallest bulb, some bulbs are made of quartz or filled with halogen. These new-technology bulbs are very expensive but save energy and need replacement less often than hotter, higher-wattage standard bulbs.

Light fixtures are sealed to be completely watertight, so the air inside reaches extreme temperatures unless cooled by contact with the water. Never turn these lights on if there is no water in the spa.

Most fixtures can be equipped with overlay lenses available in a variety of colors. These plastic lenses snap over the glass lens to change the color of the light. Being very thin plastic, these too will quickly melt if the light is operated out of the water.

Fixtures are held on the wall of the spa in a light niche. A niche is a metal can large enough to contain the fixture that is attached to the side of the spa with a watertight sealant. A waterproof conduit leads away from the niche up above water level out of the ground to a junction box for making the necessary electrical connections.

REPLACING A LIGHTBULB

RATING: EASY

The most important factor in bulb replacement is to maintain the waterproof integrity of the fixture. Follow each step carefully to avoid leaks which not only will damage the new bulb and fixture, but also may lead to electrical shock of the next person in the spa. The procedure outlined here is aimed at larger fixtures in an inground spa, but the same concepts apply to smaller fixtures in a portable spa.

1. **Preparation** Shut off the power source at the breaker. Find the junction box, also called a *J-box*. On older pool/spa combinations, this will be in the deck directly above the light niche, often under a 4-inch-diameter (10-centimeter) round stainless steel (or bronze) cover-plate, held in place by three screws. Modern installations will have the junction box at least 5 feet (up to 2 meters) from the edge of the

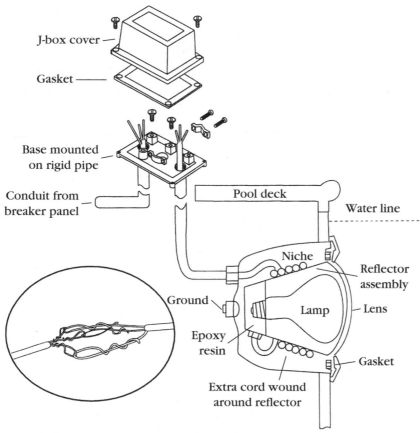

FIGURE 6-13 Exploded view of spa light fixture.

water and 18 inches (45 centimeters) above the surface of the water; so look in the garden directly behind the light niche, and you will most often find it there.

2. **Disconnect the Wiring** Figure 6-13 shows a typical junction box (J-box). Remove the four corner screws and take off the cover. Three wires will come into the box from the breaker or switch, and three wires go out of the box to the light fixture. The three to the light fixture will be individually insulated, colored white, black, and green, and are bound together with a single rubber sheath that waterproofs the package. This is the actual cord of the light fixture as described above. Disconnect the three wires. Unscrew the cord clamp.

3. **Remove the Fixture** Lean into the spa and remove the face rim lockscrew from the faceplate that holds the fixture to the niche.

The top of the fixture will float outward; the bottom is hooked into the niche and can be simply lifted out. Uncoil the excess cord to give you enough slack to raise the fixture out of the water onto the deck.

4. **Disassemble the Fixture** Remove the lens clamp assembly (Fig. 6-13), and gently pry the lens away from the fixture, taking care not to gouge the lens gasket.

5. **Replace the Bulb** Inside of some fixtures you will find a bare coiled spring wire. This is nonelectrical but is designed to break a circuit. Notice that without a bulb in place, the spring lays to one side of the fixture. Hold it up against the opposite side and screw in the new bulb. The spring lays on the bulb itself. If the bulb bursts when in use, the spring sweeps across the filament, cutting the electricity in the circuit. In this way, if water has gotten into the fixture, a live electrical circuit won't stay in contact with the water, potentially electrocuting someone in the spa.

6. **Test the Bulb** Now lay the fixture on the deck and turn on the light to make sure the new bulb works. Never operate a closed fixture without the water against it to keep it cool; but with the lens off, the heat can escape without problem.

7. **Reassemble the Fixture** I always reassemble a light fixture with a new gasket. After long use in the presence of heat and harsh chemicals, the old one is probably compressed, and if it doesn't fail immediately, it may fail before the next bulb change.

8. **Test the Assembly** Lay the fixture into the water. Hold it underwater for several minutes to make sure it doesn't leak. A few bubbles may rise from air trapped under the lip of the clamp or faceplate, but a steady stream means the fixture is filling with water. Take it apart again and dry it thoroughly. Go back to step 7 and be more careful with the reassembly. If it passes the leak test, turn it on for a few seconds before putting the fixture back in the niche. Now reset the fixture in the niche by reversing the procedure outlined above.

REPLACING A SPA LIGHT FIXTURE

RATING: ADVANCED

The most common spa light problem is simply a burned out bulb, but you may need to replace the entire fixture when it rusts out or otherwise fails. The following procedure applies to inground spas with extended wiring that is connected via a J-box to the power supply. On portable spas, the process will be similar, but the wiring will be connected to a terminal block in the spa cabinet and replacement will be self-explanatory.

1. **Prepare** Shut off the power source at the breaker. Find the junction box as described above (see "Replacing a Lightbulb").

2. **Disconnect the Wiring** Using Fig. 6-13 as a guide, follow the procedure described above (see "Replacing a Lightbulb").

3. **Remove the Fixture** Lean into the spa and remove the face rim lock-screw from the faceplate that holds the fixture to the niche. The top of the fixture will float outward; the bottom is hooked into the niche and can be simply lifted out. Uncoil the excess cord to give you enough slack to raise the fixture out of the water onto the deck. Cut the cord where it attaches to the old fixture. Strip back the rubber sheath about 6 inches. Remove the string or paper threads that also run alongside the wires (these were put in the cord to add strength). Remove 6 inches (15 centimeters) of insulation from each wire.

4. **Prepare the New Fixture** Float the new fixture in the spa. I have laid the new fixture on the deck, and it always gets tugged off or kicked, shattering the lens. So just get in the habit of laying it in the water, and you won't face this problem. Take the wires of the new fixture and strip the cord back as described in step 3. Now bind the wires of the new cord to those of the old cord, folding each wire over as if to make a hook (Fig. 6-13 inset). Now use electrical tape to cover the exposed wire, wrapping it tightly and thoroughly. Don't tape so much that you make a connection thicker than the cord itself. It won't pass through the conduit easily. The idea is to make a union of the wires that will not separate when you pull.

5. **Pull the New Cord through the Conduit** Lay the new cord out freely into the water. Pull the old cord at the J-box until you have pulled the connection and the new cord through. Take up most of the slack in the cord, but leave enough cord on the spa side so that the fixture can still be lifted out of the spa for future maintenance. Now untape your connection and discard the old wire.

6. **Install the New Fixture** Reach into the spa and coil the excess line around the fixture, and reset the fixture in the niche. Look for excess bubbles that could be a sign the fixture is leaking. Leaks don't often occur in new fixtures, but are not impossible. If there

TRICKS OF THE TRADE: LIGHT SAFETY

Electricity and water mix only too well, often with deadly results. Therefore, take precautions:

- If a fixture, lens, or gasket looks suspicious, replace it. It's not worth the time, hazard, or money to "try that old one, one more time."

- Use the bulb and gasket or replacement lens that fits the fixture. Try to get the same manufacturer's replacement parts, or use the generic after-market brand designed for that make and model. You may force a bulb, gasket, or lens into place, and it may stay water-tight for a few days. But what happens when it ultimately overheats and leaks, bringing water, swimmer, and electricity into contact?

- Don't put higher-wattage bulbs into a fixture than it was designed to take. If it isn't marked and you can't read the old bulb, don't use greater than 400 watts in your replacement.

- Even when changing a bulb, as with any repair, you are required to work to local building and health department codes. This may mean bringing an older installation up to code by adding a ground fault circuit interrupter (GFI).

appears to be a leak, disassemble and dry all components (as described above in "Replacing a Lightbulb"). Reassemble and tighten all components and test again. Finally, replace the lockscrew on the top of the niche to secure the new fixture in place.

7. **Connect the Wiring** Cut off any excess cord at the J-box, leaving enough to make a good connection with the electrical supply wiring. Reconnect the wiring and close the J-box. Turn on the light to test the new fixture.

Fiber Optic Lights

The most significant development in water lighting over the past two decades is the use of fiber optics (Fig. 6-14). The hardware looks the same, but instead of running electricity to a fixture in the spa, the cord in the conduit contains thin plastic fibers that conduct light, not electricity. The fibers terminate in a lens in the water to shed the light into the spa, or the illuminated cable is simply mounted in the spa to create a line of light all around the spa. The light source (called the *light generator*) and electricity source are located safely away from the body of water.

Materials used today are acrylic fibers bundled in multiple-strand cables instead of the old-style solid-core fiber cables. The solid-core style actually transmits more light, but is more sensitive to moisture and hardens over time, whereas the multi-strand cable is more flexible and less prone to failure from intrusion by moisture. There is a benefit of solid-core fiber cable, though—multiple sections of the solid-core fiber cable can be spliced end to end as needed, while it would not be practical to do that with multistrand cables.

Installation and troubleshooting of fiber optics are unique to each manufacturer and are probably better left to a pro. That said, if you are comfortable with electrical

FIGURE 6-14 Spa illuminated by fiber optics. *Fiber Optics Technologies.*

installations and plumbing techniques (running fiber cable to the spa through conduit is very similar to plumbing procedures), you may be comfortable installing your own fiber optic lighting. Read the manufacturer's instructions carefully and start with a few basic guidelines for the project, as described in the adjacent "Tricks of the Trade: Fiber Optics."

TRICKS OF THE TRADE: FIBER OPTICS

RATING: ADVANCED

The following dos and don'ts are provided to assist you in better understanding both the system and the installer, if you choose to use these systems.

- When you are choosing the system for your spa, don't skimp. More strands per cable and a larger light generator are more costly, but will yield superior results.

- In large spas, install several smaller lights for better results instead of relying on a single larger one.

- Shorter runs of fiber cable (from the light generator) result in less loss of light and a less costly installation.

- Cuts are critical: you may be cutting 50 to 400 individual strands within a bundle, and each one can be scratched if cut wrong, resulting in light loss. Never use knife blades or mechanical cutters.

- Locate the light generator above the water level of the spa, and ensure that it stays waterproof.

- Do not bend the final 12 inches (30 centimeters) of cable before the lens. The amount of light produced will be severely reduced.

FIGURE 6-15 Solar-powered spa light. *Poolmaster, Inc.*

Solar-Powered Spa Lighting

Poolmaster, a company that makes a variety of pool and spa accessories, has developed a unique new solar-powered floating light (Fig. 6-15). A light sensor automatically turns the light on for 6 hours at night, while rays from the sun recharge the unit's battery by day. The solar-powered light will illuminate any spa quite nicely and doesn't take up much space, measuring under 6 inches (15 centimeters) in diameter and 3 inches (7.5 centimeters) in height.

Air Blowers

An electric-powered mechanical device that forces air into the spa is called an *air blower* (Figure 6-16). Air blowers are rated by horsepower—1, 1.5, or 2—and are available as above-ground units or burial blowers, designed to be buried in the ground to reduce operating noise. Blower casings are generally made from plastic or metal.

Blowers are centrifugal air pumps, just as the circulating pump is a centrifugal water pump. The motor is the same design used on vacuum cleaners. Air is pulled in from the outside, drawn over the motor to keep it cool, then forced out to the spa. Air is pulled into the blower from impellers, again much like a water pump. Larger units use two impellers to draw more air. It is this air suction that creates the noise associated with blowers, not the motor itself.

The air blower adds general turbulence to spa waters for a general massaging effect, or it is plumbed into the air line of spa jets to turbocharge the jet massage. A blower can also aerate an air ring, a length of 2-inch (50-mm) flex PVC pipe with the holes drilled into it, laid at the bottom (or under the floor) of a spa, which distributes bubbles throughout the water column. In either case, selecting the correct size for the installation is important. If the blower is too large, it will try to push more air than the system can handle and the blower motor will overheat and burn out. If it is too small, the air delivered will be weak.

When replacing an existing blower, you can generally replace it with one of the same size. This is, however, a good opportunity to make sure the original unit was sized correctly. Air blower sizing is calculated on the depth of water through which the air must be forced. Water has weight, and the greater the depth, the greater the weight. With this increasing weight, the blower requires increasing force to push air through the water. For easier calculations, the resistance created by horizontal plumbing and angled fittings is expressed as depth so you can determine the total depth of the water that the blower must overcome. To calculate the size of the blower required, proceed as follows.

1. **Depth** Measure the actual depth of the water above the air outlet. This is not the depth of the spa, unless the air outlet is a bubble ring in the floor. It might be the depth of the jets if the blower is turbocharging the jets; it might be the depth of the seat if the bubble ring is in the seat. Remember that water displacement by bathers will raise the water level several inches.

2. **Pipe** Blowers are plumbed with 2-inch (50-millimeter) PVC. Measure the total length of all pipe between the blower and the spa. For every 10 feet (3 meters) of pipe, add 1 inch (25 millimeters) to the total water depth.

3. **Fittings** For each 90-degree angle (one 90-degree fitting or two 45-degree fittings), add ½ inch (13 millimeters) to the total water depth.

4. **Calculations** Add the three components to arrive at a total water depth. If it is less than 36 inches (90 centimeters), a

A

B

FIGURE 6-16 (A) Typical side-mounted air blower. (B) Typical bottom-mounted blower.

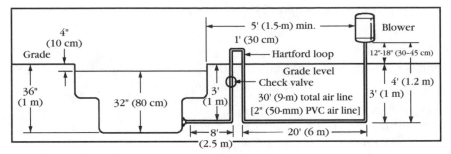

FIGURE 6-17 Plumbing installation of air blower.

1-horsepower blower is required; from 36 to 48 inches (90 to 120 centimeters), a 1.5-horsepower blower is required; for 48 to 55 inches (120 to 140 centimeters), a 2-horsepower blower is required.

Figure 6-17 depicts a typical above-ground blower installation.

Water depth = 32.0 inches (80 centimeters)

Total pipe length of 41 feet ÷ 10 = 4.1 inches (10 centimeters)

Five 90s × 0.5 inch = 2.5 inches (6 centimeters)

Total depth of water = 38.6 inches (96 centimeters)

Rounded to the nearest inch means this example is 39 inches (1 meter) of total water depth, requiring a 1.5-horsepower blower. As noted above, you don't want to oversize the blower, but when your calculation is right on the edge, move up to the next-higher unit.

5. **Air Holes** Make sure that your system can handle the volume of air. If the blower is serving an air ring, use the following calculations for the number of holes and their respective size, regardless of horsepower. Of course, if you want fewer holes, each of which is supplying more air, use a ¼-inch (6-millimeter) hole. If the spa is large and you want an even distribution of many smaller holes, use the ⅛-inch (3-millimeter) holes or something in between for your installation.

If the air ring has

⅛-inch (3-millimeter) holes, drill 96 of them

5⁄32-inch (4-millimeter) holes, drill 73 of them

3⁄16-inch (5-millimeter) holes, drill 50 of them

¼-inch (6-millimeter) holes, drill 30 of them

If the ring has too many holes or they are too large, you can plug some with plaster (on plaster spas) or silicone sealant. If the holes are too small or there are not enough, select the proper drill bit size and enlarge the existing holes and/or drill new ones. As noted previously, if the blower is supplying standard 1½-inch (40-millimeter) spa jets, make sure there is one jet per ¼ horsepower of blower.

Installation

RATING: ADVANCED

1. **Plumbing** Now that you have selected the proper blower, the installation is critical to peak performance. To prevent water from entering the electric motor, install the blower higher than the water level and plumb in a Hartford loop (Fig. 6-17) and a check valve. The loop must be 12 to 18 inches (30 to 45 centimeters) above the water level to be effective. When reducing from the 2-inch (50-millimeter) pipe of the blower to the 1½-inch (40-millimeter) jet plumbing, make the reduction at the T or three-port valve, not near the blower. A check valve is also recommended in the system of not more than ½ pounds per square inch (34 millibars) pressure. If you use the loop and the antisurge valve provided with most blowers, the check valve is one restriction in the system too many. The other function of the check valve and/or loop is to keep the water as close to the spa as possible and fill the lines from the blower with air as much as possible. The result is that the air reaches the spa faster than if it must force itself through water in the pipes before forcing itself through the water in the spa. Therefore, the loop and check valve should be located as near the spa as possible. Antisurge valves and check valves built in at the blower are for added protection.

2. **Mounting** Blowers are provided as side discharge, bottom discharge, or convertible. Figure 6-17 depicts a bottom discharge, while a side discharge is the same product but with the air discharge on the lower portion of one side. The convertible type has discharge ports on both the bottom and side, and a plug is used to close off whichever discharge port is not needed. The plug is either screwed in place or attached with some silicone sealant. I have found that on plugs that are not designed to be screwed in place, you can touch a dab of PVC glue to one side of the plug and insert it. Again, give the fumes adequate time to dissipate before turning on the unit for the first time.

TRICKS OF THE TRADE: BLOWER INSTALLATIONS

- When designing the plumbing, use two 45-degree angles to make pipe turns rather than one 90. As described in Chap. 3 on pumps, less resistance is created in this way.

- When gluing the plumbing together, allow the fumes to dissipate by waiting at least 24 hours before operating the blower. An electric spark can ignite the fumes. Most manufacturers recommend dry-fitting the blower to the stub-out pipe rather than gluing.

- Silicone sealant will keep moisture out of this connection and hold the blower in place. Over time, a dry-fitted blower without silicone or glue will vibrate loose and literally blow off the plumbing.

- In addition to the check valve and Hartford loop, some manufacturers provide a small rubber flapper insert called an *antisurge valve* that is inserted in the discharge port before mounting it to the pipe. This check valve provides security that the air can exit the unit but nothing else will enter. By the way, because the antisurge valve is not listed by Underwriters Laboratories (UL), it cannot technically be called a check valve, thus the alternate term, *antisurge valve*. Some manufacturers build the check valve inside the discharge pipe of the unit, so look inside to determine if you need one.

- Always read the manufacturer's literature provided with the blower.

Burial blowers are installed in the same way, but the unit is buried for noise reduction, usually no more than 12 inches (30 centimeters) deep. An air vent stack runs above ground for ventilation. Otherwise the sizing and plumbing recommendations are identical to those of the above-ground unit.

Because the blower vibrates somewhat, be sure it is secured. In Fig. 6-17, I would drive a stake or length of rebar into the ground and use a plumber's strap to secure the pipe to it. If the deck is concrete, bring the blower pipe up adjacent to a wall so the blower can be strapped to the wall. If the blower is a side-discharge unit, it can be mounted to a shelf or other surface directly.

3. **Electrical** Finally, the blower must be connected to the electrical supply. Blowers are supplied with either 110-volt or 220-volt motors. Unlike pump motors, these are not convertible, so be sure to buy the unit that corresponds to the equipment electrical supply. Some blowers are supplied with a built-in switch; but whether

they have one or not, a wiring J-box is mounted to the unit for connection of the conduit (use waterproof conduit for outdoor installations) and wiring from the electrical supply.

Each unit includes a ground wire for bonding to the electrical panel's ground bar, but if the unit is installed within 10 feet (3 meters) of the spa, additional grounding must be used. A bonding wire terminal is provided on each blower to directly run a #8 copper grounding wire into the ground for this purpose or for attachment to the bonding system of the other equipment. Consult with your electrical subcontractor and/or follow local building codes. Also follow the recommendations and wiring diagram provided by the manufacturer.

For the blower to operate correctly, it must be provided with adequate amperage. A 2-horsepower blower will draw up to 13 amps at 110 volts, meaning the electrical service required is substantial. If you are having performance problems with a correctly sized blower, check the amperage to be sure it is supplying that required by the blower. The amperage is listed on the rating plate of the blower, along with the manufacturer's name, voltage, and horsepower. Sometimes the horsepower rating is obscured (on some manufacturers it is not printed on the rating plate). Since more horsepower needs more amperage, you can use this general rule of thumb to estimate the horsepower of the existing blower:

- 1 horsepower: 6.5 amps at 110 volts, 3.25 amps at 220 volts
- 2 horsepower: 13 amps at 110 volts, 6.5 amps at 220 volts

Blower Repairs

RATING: EASY

The motor, brushes, and other components of a blower can be replaced, but an entire unit costs less than $100 for most sizes. Since the labor and parts for a repair can cost that much, I have never recommended fixing a blower to a customer unless the unit is reasonably new.

Most blower failure is caused by overheating, which in turn is caused by too much restriction on the system. In other words, before replacing a fairly new blower, check the sizing carefully. Blowers heat up even when used in a well-designed system, so failure is inevitable. A blower should last 3 to 5 years in outdoor installations with regular use, and twice that time for indoor installations.

Some blowers are built with a thermal overload switch similar to that on pump motors. When the operating temperature exceeds the designed amount, the switch opens and the blower shuts off. When the air cools, the switch cools as well, closing it and restoring electric current to the motor. If your blower cuts on and off, it might be a sign of inadequate ventilation around the blower (often), system pressure creating an overheated motor (most often), or a faulty thermal switch (rarely).

Skidpacks

A skidpack (Fig. 6-18A) refers to a combination of spa equipment mounted on a metal frame called a skid. The pump/motor (usually two-speed), heater, blower, and control devices are all combined on the skidpack for compact installations, mostly on portable spas. Filters are sometimes included, but more often are mounted in-line in the plumbing (Fig. 6-18B) or built into the wall of the spa itself (refer to Fig. 4-1B).

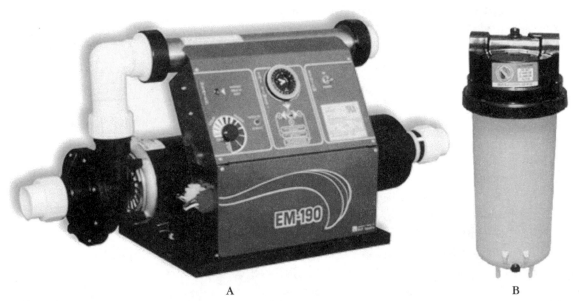

A   B

FIGURE 6-18   (A) Portable spa equipment skidpack. (B) Portable spa in-line filter. (C) Portable spa waterfall.

Some portable spas feature a small circulation pump that discharges to a waterfall feature (Fig. 6-18C).

Most skidpacks are designed to be plugged into a wall outlet in the home at 110 volts, but larger ones which require greater amperage are designed for 220 volts supplied by a dedicated circuit from the breaker panel. Some skidpacks are convertible for either voltage.

Components of the skidpack are susceptible to the same maintenance and repair problems as individual components. The difference is that to repair a component of a skidpack, you may have to disassemble several other components from the pack to access the one needing repair.

C

FIGURE 6-18 *(Continued)*

The second major challenge of skidpacks is the control system. There are far too many designs to profile a "common" control system for skidpacks, since they may include any or all of time clocks, air switches, various thermostat controls, light controls, and high/low pump speed controls. Some manufacturers provide a plug-in circuit board, where each component plugs into an outlet designated for that item. The switching system controls electrical flow to each outlet. When a blower fails to operate, for example, you can literally unplug the blower and test the outlet when the switch for the blower is on. If there is no current, you know the problem is in the switch, wiring, or other control circuit rather than the blower itself. If there is current, then the blower can be easily removed and replaced. Testing for current can be done with your multitester or a special plug-in tester provided by the manufacturer. Since many skidpack makers have come and gone from the business, you may have no choice but to fool around with the controls or tear them out and replace them with a more reliable air switch system. Finally, if the skidpack is fairly old, it may not make sense to replace expensive components when an entire skidpack can cost as little as $500.

If the skidpack is operating on a 110-volt system, sometimes performance problems can be traced to inadequate amperage. Check the electrical supply circuit if the blower appears to be running slowly or the motor overheats and shuts off frequently. Call in an electrical subcontractor if the circuit needs upgrading. As described previously in this chapter, portable spas should be protected with a ground fault circuit interrupter, either at the electrical wall outlet or as part of the skidpack plug itself.

Finally, as with any spa equipment repair, make careful note of how the skidpack comes apart, to make reassembly easier. This point is extremely important with skidpacks, since they often require so much disassembly to get at a particular component and since so many manufacturers are no longer in business to offer technical advice.

Unique Spa Accessories

No other part of the pool and spa industry has seen such innovation in recent years as the accessories now available for a spa or hot tub. From simple pillows and cushions to make relaxing in the spa more comfortable, to audio and video equipment that can cost thousands of dollars and turn your spa into a living room, spa accessories are now as diverse and creative as your imagination.

Audio and Video Systems

Figure 6-19A shows a portable spa with both a TV/video monitor and a four-speaker stereo audio system (speakers mounted at each corner). Larger spas are now equipped with retractable plasma screen TVs that measure as much as 4 feet (1.2 meters) diagonally. Add a larger waterfall, special hydrotherapy nooks, special wooden towel/robe warmers, and a dining area (yes, a dining area in the spa!) and you can easily spend over $30,000 for your "portable" spa.

Audio (Fig. 6-19B) and video systems bring complex electronics within range of water and harsh chemicals, so installation of these devices is best left to the pros. You might find it less costly to trade up to a spa which already includes these features and is covered by the manufacturer's warranty, than to remodel your existing unit.

Steps, Rails, and Seating Accessories

Steps that help you get in and out of your spa are available in every size, color, and material. Figure 6-20A shows steps and accessories made of plastic, but finished to look like wood. Be careful, when you select a step unit, to choose one with nonskid molded in to the plastic surface or added to wooden units. Many people use bath oils in spas, making them more prone to slips and falls when climbing in or out. Generic bolt-on handrails are also a good safety accessory (Fig. 6-20B).

Neck and Booster Seat Pillows

Another practical accessory for any spa is a simple neck pillow, which makes lounging in the spa more comfortable. Figure 6-21 shows a typical spa pillow that is preset to accommodate a bather using the massage jets. Other pillows have brackets that allow them to hang over the edge of the spa and are easily adapted to any make or model. Each spa manufacturer also sells pillows designed specifically for its models, so you may want to check websites before ordering the one that's right for your spa. Pillows should be replaced every year, since mold and mildew will

A

FIGURE 6-19 (A) Spa TV/video/audio system. (B) Spa stereo system. *A: Master Spas.*

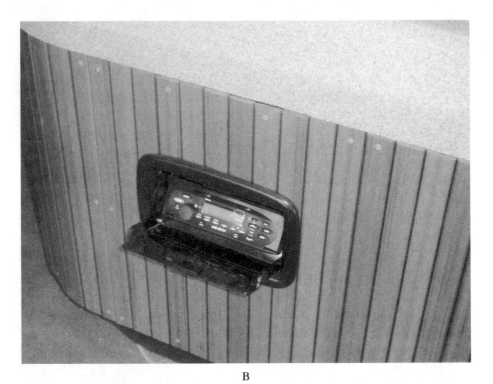

B

FIGURE 6-19 *(Continued)*

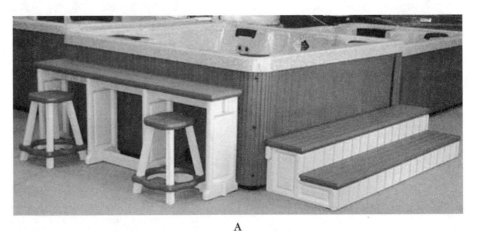

A

FIGURE 6-20 (A) Spa steps, bar, and seating accessories. (B) Wood step and metal bolt-on hand rail. *A: Confer Plastics.*

accumulate in and around the fabric, no matter how clean you keep your spa.

Booster seat pillows are another great innovation (Fig. 6-22). Not all bathers are the same size, so you may find that the seats in your spa are just not fitted for your body size or contour. The booster seat pillow not only helps you adjust to the seat height, but also can help position your body near therapeutic jets, wherever your aching muscles are located. Finally, booster seat pillows make sitting for long periods on hard plastic or wood much more enjoyable.

Sun Shades and Gazebos

A popular addition to the backyard spa is a sun shade (Fig. 6-23) or gazebo (Fig. 6-24). Sun shades are made by numerous manufacturers, some of which also make spas, and come in every size and color you can imagine. Ask your spa dealer or surf the Web for the one that fits your needs and pocketbook.

A gazebo will add comfort to your spa by providing shade and some protection from rain or snow. There are many makes and models, some designed specifically for particular spa brands, but there are also plans and kits that take you through the process of building your own gazebo. Again, some research on the Web will help you find the one that's right for you. Check "Spa and Hot Tub Resource Guide" at the back of the book for helpful websites.

Sprays

Since the bather sits in the hot water of a spa, many people enjoy a spray of cool water occasionally. This is something that can be inexpensively added to any spa.

B

FIGURE 6-20 (*Continued*)

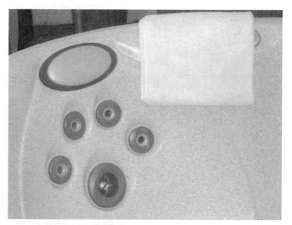

FIGURE 6-21 Spa neck pillow.

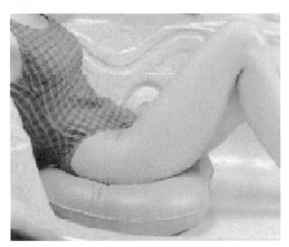

FIGURE 6-22 Spa booster seat pillow. *Essentials.*   **FIGURE 6-23** Spa sun shade.

FIGURE 6-24 Gazebo for portable spa. *Gordon & Grant Hot Tubs.*

The spray is supplied with water from a garden hose. Run the hose to the edge of the spa and plumb it to a ¾-inch (2-centimeter) PVC pipe mounted about 3 feet above the spa. Secure the pipe to a wall or stake in the ground or deck. Plumb a suitable nozzle onto the end of the ¾-inch (2 centimeter) pipe to create a mist of cool water.

Some showerheads have mist settings, but I use an actual bronze garden mist head, available at any garden supply store for a few dollars. To give bathers control over the mister, install a ball valve between the garden hose and the ¾-inch (2-centimeter) pipe at a convenient location on the lip of the spa.

Thermometers

Some commercial pools require that an in-water thermometer be available at each spa, and most residential spa owners want one, too. For commercial installations, I recommend the unit that is built into a skimmer cover (Fig. 6-25A) and takes the temperature at the skimmer. Other models usually disappear or are tampered with.

You can also simply tie a thermometer to a rail or ladder with the tube-type model that has a string attached for this purpose, or you can provide a tube model that has a float on the top and simply floats on the surface of the water. These usually float (Fig. 6-25B) into the skimmer basket, which is a good place for them to keep out of sight.

For you high-tech techies, digital-readout, battery-operated thermometers are also available in floating models or with probes that you put in the water while reading the information on a small handheld device about the size of a calculator.

Scents

Dozens of products are designed to provide a wide variety of pleasant odors to your spa water. Most are harmless, but some contain oils that are designed to make the water feel "silky" but actually hasten the clogging of the filter and should be avoided. Scents are a matter of personal taste, but one word of caution—use them sparingly at first. Much like adding salt to soup, you won't notice the result at first and may be tempted to use a heavier hand. After several minutes of heat and bubbling, the aroma will intensify and you may be sorry you used the product at all. Moreover, scents will linger, even after you drain and clean the spa, so be sure any scents that you use are ones you are prepared to live with for awhile.

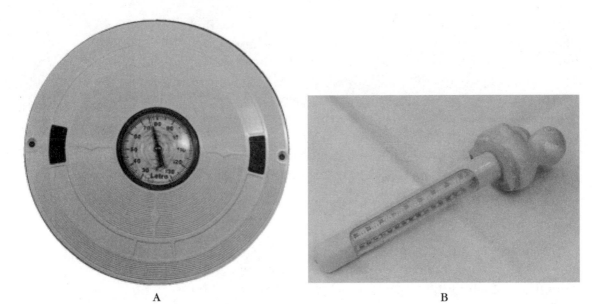

A   B

FIGURE 6-25   (A) Skimmer lid thermometer. (B) Floating thermometer.

Safety Signs

Most building or health department codes require certain signage to be prominently displayed near the spa. Check your local codes for specifics and required proximity to the water's edge. The basic signs (Fig. 6-26) you need (some or all, depending on your local codes) are as follows:

- **Maximum Bather Load** This sign usually says, in simple bold, block letters, "Occupant Capacity _____" or "Maximum Occupants _____." You fill in the blanks, depending on the local codes. Many jurisdictions allow 15 square feet (1.4 square meters) of surface area for each bather, so a 15-foot × 30-foot (4.5-meter × 9-meter) pool would allow 30 maximum occupants at a time (450 square feet of surface area ÷ 15 = 30). Other rules guide spa occupancy or wading pools for kids, so check local requirements before you write the number in the blank provided. Use paint or an indelible marker so the number cannot be changed by pranksters.

- **Artificial Respiration** This sign diagrams artificial respiration and first aid techniques in case of emergency. It's a good idea to familiarize yourself with these techniques in case of emergency.

- **Emergency Phone Numbers** This sign prominently displays 911 for general emergency calling and also provides spaces to include the phone numbers of a local emergency room or doctor, fire department, and police. Since most communities now have the 911 emergency number, some jurisdictions only require this to be posted.

FIGURE 6-26 Safety signage for spa.

- **Pool/Spa Rules** This sign is available with residential or commercial rules and includes hours for use, age restrictions of users, commonsense rules such as "No glass" or "No running" allowed on decks, etc. These rules may be generic or may have specific language required by your state.

- **Emergency Shut-off** This sign is posted next to an electrical switch that cuts all power to all equipment in case of emergency. It simply makes users aware that the shut-off exists in case someone is injured in the spa or, more often, when clothes get sucked into powerful pump suction openings in the spa.

Covers

Covers are an important spa accessory to keep the water clean, to save heating energy costs, and to reduce evaporation. Shielding the water from the sun further cuts down on chemical use. There are more choices of covers than there are makes and models of spas themselves, which gives you good options depending on your particular needs and budget.

Covers can also provide safety. Heavy-duty spa covers can prevent kids from falling into spas, especially when combined with straps or locks to ensure the covers can be removed only by adults. Of course, not all covers are built with these properties, so the intended users are a critical factor in selecting the right spa cover.

Bubble Solar Covers

Have you ever unpacked something sent in the mail that is wrapped in that plastic bubble wrap? Someone discovered that large sheets of this material make good spa covers.

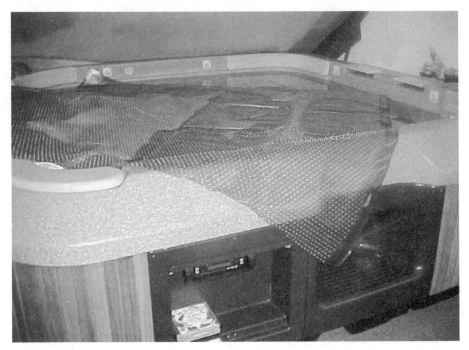

FIGURE 6-27 Bubble cover.

Figure 6-27 shows a typical bubble cover (also called *sealed-air* cover for obvious reasons). In profile, the cover has one flat side and one bumpy side. In fact, the cover is made from two sheets of blue plastic (usually available in 8- or 12-millimeter thickness), heat-welded together with air bubbles in between.

The sun warms the air bubbles, which transfer the heat to the water (thus the term *solar* cover). Similarly, the trapped air acts as an insulator for the heat coming up from the water. Always lay a bubble cover on the water with the bubble side down. In this way, the spaces between the bubbles also act as pockets for trapped air, further insulating the water.

Because they are thin, lightweight, and flexible, bubble covers can be easily cut to any size with a scissors or razor knife and are sold in large sheets of almost any size. Bubble covers are quite inexpensive and last 2 to 4 years, depending on water chemistry, weather conditions, and user wear-and-tear.

The disadvantage of a bubble cover is that in heavy winds, it can blow off, and it doesn't really keep out dirt or debris, because as you remove the cover, the dirt falls into the spa anyway. Also, as the bubble cover ages, sunlight and water chemicals make the plastic brittle and the bubbles collapse, sending little bits of blue plastic into the spa and circulation system. In short, bubble covers are only good for their thermal properties, which are valuable, especially if you heat the spa year-round.

INSTALLATION

RATING: EASY

As you might guess, installation of bubble covers is very user-friendly.

1. **Measure** If your spa is a standard rectangle or circle, measuring is quite simple. Be sure to buy a cover that will completely cover all the water surface. In general, you are better off to buy a size larger than you actually need, to ensure a complete fit. If your spa is an irregular shape, take a sheet of clear heavy plastic (available in hardware stores) and lay it across the top of the spa or hot tub (Fig. 6-28), taping it securely. Use a permanent felt-tip marker to draw the outline of the water and an outline of the perimeter of the spa. Buy a cover that is slightly larger than the largest measurement on your template.

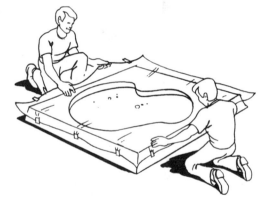

2. **Prepare the Cover** Lay the cover out over the water surface, and leave it for 2 or 3 days. (Some manufacturers recommend as many as 10 days, but unless it is very cold, I have not found any difference after 2 or 3 days.) You can remove it to use the spa during this period; however, the idea is to give the material time to relax to its full size in the sun and to shrink to any degree that it might (I have found shrinkage is not more than 5% in any direction).

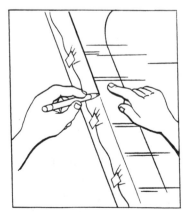

3. **Cut to Size** Using a shears, scissors, or razor knife, cut the cover to the water surface size

FIGURE 6-28 Measuring for spa cover.
Sunstar Enterprises.

of the spa. As you walk around the cover, cut slightly less than you think you should—you can always go around again and trim off a bit more, but it sure is hard to add any back on if you cut too much! If you have an irregular-shape spa and have made a template in step 1, cut out the plastic template to the shape of the spa and tape it to the bubble cover, then cut the cover to size.

Foam

Much like bubble covers, foam covers are sheets of lightweight compressed foam [⅛ inch (3 millimeters) or more thick] that float on the surface of the spa (Fig. 6-29). Foam is about 6 to 10 times more expensive than bubble plastic, but has greater insulation benefits and will pay for itself in no time. Installation is the same as that for a bubble cover, described above.

Custom Covers

Many manufacturers can custom-make a cover for a spa, one that either floats on the surface of the water or covers the entire unit. Providing careful measurements is the key to a good fit and best insulation properties.

Measure your spa for a custom cover, using the same sheet plastic template method described above for bubble solar covers. For a custom cover, make any notations of importance directly on the plastic, such as the manufacturer of the spa and the model. Since most covers are "hinged" in the middle, indicate preferences regarding placement and the number of hinged segments you desire. Custom covers come in a wide variety of materials and design, depending on your needs, but all will be fairly expensive.

WOOD

Any good carpenter can make a wooden cover for a spa that sits on the deck over the top of a spa. These are particularly popular with wooden hot tubs and where child safety may be a factor. They are sturdy, and lock hardware can easily be added. Most wood covers are made in two sections for easier handling and storage. The sections can be hinged for convenience or left separate for lighter handling weights.

FIGURE 6-29 Foam cover material.

If you contract with a carpenter to make a custom spa cover, make sure he or she uses stainless or bronze nails and screws that won't rust. Check the size of the job and get an estimate from the carpenter of the weight of the materials—when a wood cover gets past a 6-foot diameter, it may be too heavy for the customer to handle. Also, make sure redwood, cedar, or other hardwoods are used. Pine or fir will rot quickly unless it is varnished. Don't paint wood covers—the heat and chemicals of the spa water evaporating will peel paint very rapidly.

WOOD AND FOAM

One variation of covers we have already discussed is the popular foam cover with redwood slats, shown in Fig. 6-30. The foam connects the wood slats and provides insulation. The wood slats provide rigid safety, but obviously the combination is not as heavy as a solid wood cover. The added convenience is the ability for roll-up removal and storage.

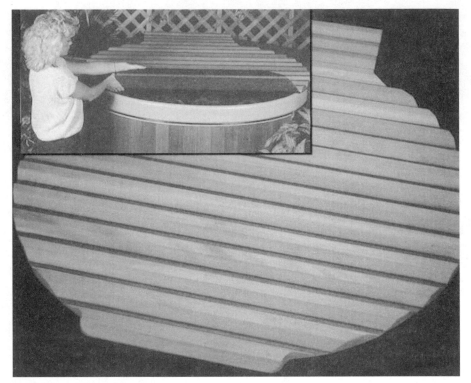

FIGURE 6-30 Foam and wood cover. *Great Northern Hot Tubs.*

REINFORCED UPHOLSTERED FOAM

A popular alternative to heavy wood is a rigid cover (that also sits on the deck rather than actually on the surface of the water) made of Styrofoam, framed for strength with plastic or aluminum and covered with plastic or Naugahyde (Fig. 6-31). These covers are usually made in two sections with a fabric hinge to connect them.

Safety Barriers

Many jurisdictions now require some type of fence, solid cover, or other safety barrier around a pool or spa, to prevent drowning. If more people paid attention to this commonsense requirement, many needless deaths and injuries could be prevented every year. Generally, such "barrier codes" include at least these requirements:

TRICKS OF THE TRADE: COVER ACCESSORIES

As spas become larger and more complex, so do covers. As a result, various manufacturers have added new components to make covers easier to use and to increase their safety value.

- Lifting units (Fig. 6-32) make it easier to remove and replace large, heavy covers, making it more likely that the cover will be used in the first place. Some operate with hydraulic lifts, much as the hatchback of a car does.

- Adjustable hold-down straps (Fig. 6-33) can be easily added to covers, providing protection against wind and added safety around children and pets.

- Special cleaning products are made by or for cover manufacturers. Don't use household cleaners or pool/spa chlorine, as most vinyls will discolor.

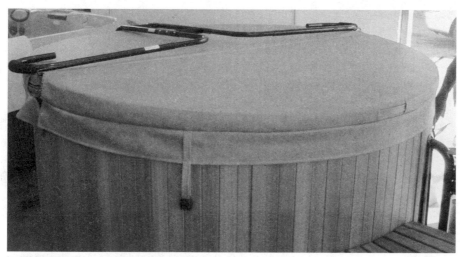

FIGURE 6-31 Custom cover.

- The spa must have a fence on all four sides at least 4 feet (1.2 meters) high.

- Gates must be self-closing and latching.

- Spas must have a *safety cover* (as defined by various certifying agencies).

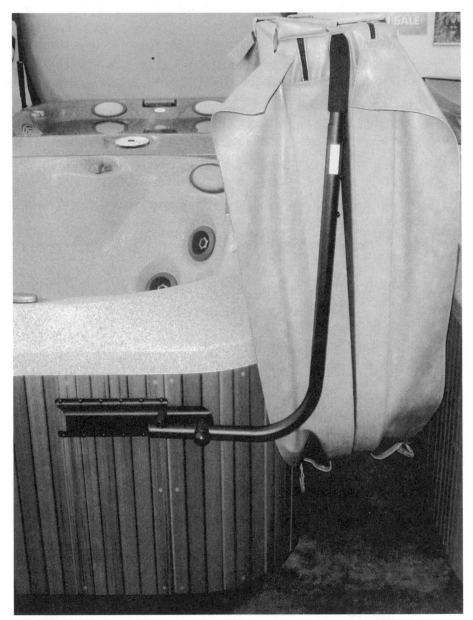

FIGURE 6-32 Spa cover lifting device.

■ Where the home itself forms one or more "walls" of the fence, doors must have locks or other provisions to prevent children from accessing the spa.

■ Sometimes an alarm can be substituted for certain barrier requirements (check on the type of alarm permitted in your area).

Remember, many drownings of children occur even when barriers are provided if gates are left open or covers left off. Safety barriers are only as good as their actual use.

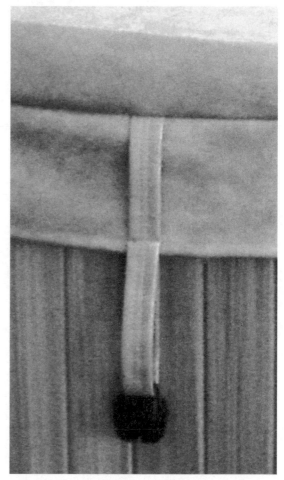

FIGURE 6-33 Spa cover straps.

Water Chemistry

et's start with the components of "healthy" water, but don't despair if this looks too technical for you to master. As the descriptions that follow will demonstrate, anyone can maintain a clean, healthy spa with a little knowledge and a very small investment of time.

- Chlorine residual (sanitizer): 1.0–3.0 ppm

- Total alkalinity: 80–150 ppm

- pH level: 7.4–7.6

- Hardness: 200–400 ppm

- Total dissolved solids: Less than 2000 ppm

Of course if you use an alternative to chlorine in your spa, the appropriate level of that sanitizer would be listed instead.

Spa water chemistry is a process of balance. Change one component, even to bring it into a correct range, and you may adversely affect another component, thereby adversely affecting the entire spa.

Imagine that the water quality parameters listed above are stones, each of equal size and weight, evenly distributed around the edge of a dinner plate and balanced on one finger. If one stone is doubled in weight or removed, it will change the balance, eventually making the entire plate crash to the ground. So it is with water chemistry—a balancing act, with each component working on the others.

This is the "balance" of water chemistry, but what is the "demand" part? Water is a solvent. It will dissolve and absorb any substance until it can no longer hold what it dissolves (called the *saturation point*). After this, it will "dump" the excess of what it has dissolved (called *precipitate*). Therefore, you can say that water makes "demands" on anything it comes in contact with until those demands are satisfied.

For example, if your spa water is very acidic, it will "demand" to be balanced with something alkaline. If your spa is plastered, the alkaline lime in the plaster will be dissolved into the water until that balance is achieved. When the water is no longer acidic, it will start "dumping" excess alkaline material, depositing it on tiles and inside spa equipment as well as back on the plaster as rough, uneven calcium deposits.

Therefore, successful water maintenance is the quest for balance between the various demands of each aspect of water chemistry. Let's examine those now.

Sanitizers

Before we proceed, there are two terms that you should understand as you evaluate which sanitizer is right for you and how effective it is after you have used it:

- **Sanitizer Demand** The amount of any sanitizer product (in any form) needed to kill all the bacteria present in the body of water.

- **Sanitizer Residual** The amount of sanitizer (expressed in parts per million) that is left over after demand is satisfied.

- **Sanitizer Availability** Sanitizers will "combine" with other elements present in the water, meaning some will be "locked up" or "unavailable" for oxidizing bacteria while some will be "available" for that sanitizing purpose (expressed as the residual).

The purpose of any sanitizer is to kill bacteria in the water, which is done by oxidizing organic material in a process akin to rust on metals. The most common sanitizer is still chlorine, because it is readily affordable and available in a variety of forms. That said, alternatives are growing in popularity and will also be described here.

Chlorine

Chlorine for spas is manufactured in gas (used only for high-volume commercial installations), liquid, dry granular, and tablet forms. A lengthy description of the chemistry of each type can be found *The Ultimate Pool Maintenance Manual,* but for most spa applications, here's what you need to know:

LIQUID CHLORINE (SODIUM HYPOCHLORITE)

Liquid chlorine has a high pH, around 13, and is full of minerals. In fact, you are adding 1.5 pounds (0.7 kilogram) of salt to the water for every 1 gallon (4 liters) of liquid chlorine you add, so this adds to water hardness. Liquid chlorine is supplied at approximately 12% strength; that is, 12% of the volume you buy is available sanitizer, and the remainder is water. Air, sunlight, and age will accelerate the deterioration of liquid chlorine, so keep your supplies fresh and covered. The advantages of liquid chlorine are that it is easy to use (just pour it into the body of water being treated) and goes into solution immediately, since it's already a liquid.

DRY (GRANULAR OR TABLET) CHLORINE

There are essentially two types of dry chlorine sanitizers, each sold by various brand names, so look at the list of active ingredients to determine what you are actually buying.

Calcium Hypochlorite: Available in granular or tablet form, calcium hypochlorite is unstable (meaning it quickly loses its ability to oxidize bacteria, especially in the presence of heat and sunlight) and slow-dissolving and leaves substantial sediment after the chlorine portion of the product enters solution in the water. It is 65% available chlorine, and the pH is 11.5, so it will tend to raise the pH of the water.

"Dichlor" (Sodium Dichloro-*s*-Triazinetrione):   Dichlor is very stable and about 60% available chlorine with no sediment or by-products and a slightly acidic pH of 6.8.

"Trichlor" (Trichloro-Triazinetrione):   Trichlor is the most concentrated (and therefore most expensive) form of chlorine, 90% available sanitizer, produced mostly as tablets that slowly dissolve, although dry granular trichlor is also produced. Trichlor's pH is around 3, so it is very acidic. One major advantage of trichlor is that it is slow to dissolve; so when a body of water is not serviced frequently, you can still expect chlorine to be released into the water on a regular and continuing basis.

SUMMARY SCORECARD OF CHLORINE PRODUCTS

To give you an overview of the chlorine products just discussed, here's a simple summary of their properties. They are arranged in order of relative cost, with gas being the cheapest. Note the relationship between cost and stability—the more stable you try to make this inherently unstable product, the more you must add to it and therefore the more expensive it becomes.

Product	pH	Available chlorine, %	Common form	Stability
Chlorine gas	Low	100	Gas	Very unstable
Sodium hypochlorite	13+	12.5	Liquid	Unstable
Calcium hypochlorite	11.5	65	Dry granular	Stable
Dichlor	6.8	60	Dry granular	Very stable
Trichlor	3.0	90	Tablet (or granular)	Very, very stable

To compare cost and volumes for equivalent sanitizing action:

Product	Volume	Cost
Chlorine gas	1 pound (0.45 kilogram)	N/A
Liquid chlorine	1 gallon (3.8 liters)	2 times the equivalent amount of gas
Calcium hypochlorite	24.5 ounces (0.69 kilogram)	3 times the equivalent amount of gas
Dichlor	28.5 ounces (0.81 kilogram)	3 times the equivalent amount of gas
Trichlor	18 ounces (0.51 kilogram)	5 times the equivalent amount of gas

CHLORAMINES AND AMMONIA

As you might have guessed from the name, chloramine is a combination of chlorine and ammonia. Remember that chlorine likes to combine with other elements in the water. It is this fact that makes it an effective oxidizer or killer of bacteria, as it tries to combine with the bacteria, thus killing them. However, when ammonia is present in the water, the two will combine to form chloramines.

Chloramines are very weak cleaners (weak oxidizers of bacteria) because the chlorine is "locked up" with the ammonia and therefore not available to kill bacteria. Chloramines form when ammonia is present in the water from human sweat or urine and when there is insufficient chlorine present to combine with all the ammonia and still have some left over to oxidize bacteria. The irritated eyes and odors usually associated with chlorine are caused by chloramines, not by adding too much chlorine to the water.

One final term you should know. *Break-point chlorination* is the point at which you have added enough chlorine to neutralize all chloramines, after which time the available chlorine will go back to oxidizing bacteria, instead of combining with ammonia in the water.

Superchlorination (aka *Shocking*):   Even if you maintain a sufficient residual chlorine in your spa at all times, the ammonia and other foreign matter in the water may keep your chlorine from being 100% available. That's why algae grow even though they have a high residual when you test the water. Superchlorination is, therefore, simply adding lots and lots of whatever chlorine product you use so that there is plenty in your water for the foreign matter to absorb at will, leaving enough over to oxidize bacteria and kill any algae.

You will only want to superchlorinate when you have a substantial volume of ammonia or other foreign matter in your spa to be oxidized. Since you can't really know how much is in the water that might be "locking up" your chlorine, you can guess based on how dirty the spa gets every week and how much it is used. Of course, the more swimmers there are, the more sweat and urine are likely to be present. Therefore, it's a judgment call.

Each chlorine product requires different amounts to superchlorinate. Follow the directions on the package for superchlorination recommendations with each product. Keep bathers out of the water and keep the water circulating 24 hours a day until the residual reads normal again.

Bromine

There are numerous alternatives to chlorine sanitizers, including other chemicals and high-technology solutions. Bromine is similar to chlorine and more stable than liquid chlorine at high temperatures, so many people prefer it for spas and hot tubs. To work effectively, however, bromine must have a catalyst, which is usually a small amount of chlorine (added to bromine product, so it is not detectable).

Bromine is produced in granular or tablet stick form and is applied in the same manner as dry versions of chlorine. People prefer bromine because it does not produce the characteristic odors that chlorine does, but as noted above, those odors are actually caused by a dirty spa, not by the chlorine itself.

Biguanicides

A relatively new sanitizer most familiar by the trade name *Baquacil,* *biguanicide* is a general term referring to a disinfectant polymer that more accurately goes by the name *polyhexamethylene biguanicide* (PHMB). A hydrogen peroxide product is applied as a monthly shock, and a quaternary ammonium-based supplement is needed weekly. PHMB concentrations need to be kept between 30 and 50 parts per million (ppm) and can be tested with a special test kit using reagent drops. Other water balance parameters will be the same as for any other sanitizer.

PHMB cannot be mixed with chlorine products or, for that matter, any other chemicals (except those designed as part of the package), or you will find chocolate brown colored water and stains on the plastic or plaster walls of the spa. This means you must follow package instructions very carefully to first neutralize other chemicals and then remove any metals that may be present in the water when changing over from a chlorine-treated spa to PHMB. Also PHMB reacts with household detergents and trisodium phosphate (TSP), so be careful with tile cleaners or any other cleaning products in a spa that uses PHMB products.

Ultraviolet (UV) Light

For UV light to act as a sanitizer, the spa water must pass through a device where it is exposed to a beam of ultraviolet light. Any bacteria in the water passing through are killed, but no bacteria are prevented from growing in the first place. Thus, UV light is not effective by itself, but requires the use of chlorine or another chemical sanitizer to establish a residual in the water. Many people like the fact that they can use far less chemical sanitizer if they also have a UV light system, but it is costly, typically over $700.

UV light systems are simple to install. Simply plumb them in-line after the heater and before the water flows back to the spa, and plug the unit into a standard household electrical outlet.

Ozone

Ozone has been used for decades to purify municipal drinking water and more recently for pool and spa sanitizing applications. It is very effective, but very unstable. Thus, it will react with contaminants immediately and break down, leaving little to attack any new introduction of algae or bacteria. Therefore, as with other alternative sanitizers, a residual of chlorine or other chemical sanitizer is always required for a total sanitizing package. As with UV light, however, using ozone can reduce the amount of chemicals employed.

Ozone for spas is created in one of two ways. One uses ultraviolet light (not to be confused with the ultraviolet systems described above). The benefit of the UV ozone generator is that it is cheaper than its counterpart (described next); the drawback is that it produces far less ozone.

The second method of ozone production is called *corona discharge* (CD). The CD ozone generator (Fig. 7-1) forms an electrical field (or *corona*) that converts oxygen to ozone, much like the creation of ozone by lightning. Generally, this produces 10 times more ozone for the same amount of electricity as a UV ozone generator.

Ozone has a neutral pH and generally has little impact on other parameters of the water balance. It is still necessary to shock the spa periodically, as previously described. Purchasing and installing an ozone system is easy. Several manufacturers provide models based on the size of the spa and/or bather load. Each comes with plumbing directions which are easy to follow if you comprehend the plumbing techniques presented in Chap. 2. Electrical connections are made via a standard appliance plug to a standard household electrical outlet.

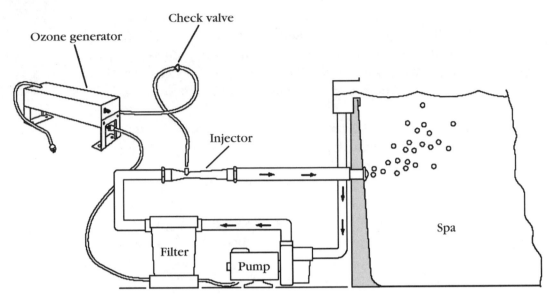

FIGURE 7-1 Corona discharge ozone unit.

pH

We have now examined the properties and types of sanitizers, but that is only the first parameter of water quality. The next component is pH.

The pH is a way to assess the relative acidity or alkalinity of water. It is a comparative logarithmic (meaning for every point up the scale, the value increases 10-fold) scale of 1 to 14, where 1 is extremely acidic and 14 is extremely alkaline. Then 7.0 is obviously in the middle and is "neutral," but for our purposes 7.4 is considered *neutral*—neither acidic nor alkaline. In fact, the Los Angeles Health Department requires a spa pH to be between 7.2 and 7.8.

The pH has an amazing effect on the water. If allowed to become very acidic (from adding too much acid, from too many bathers sweating or urinating, from too much acidic dirt and debris in the water), the water becomes corrosive, dissolving metal that it comes in contact with. If the water is allowed to become too alkaline, calcium deposits (scale) can form in plumbing, in equipment, and on spa walls or tiles.

More than that, the pH determines the effectiveness of the sanitizer. For example, you might have a chlorine residual of 3 ppm in a spa doing a fine job at a pH of 7.4. But now increase the pH just 0.4 to 7.8, and you will still have a residual reading of 3 ppm, but the ability of the chlorine to oxidize bacteria and algae will decrease by as much as one-third! Therefore, maintaining a proper pH is not just a factor of interesting chemistry or saving your equipment, it is a matter of effective maintenance.

The pH is adjusted with various forms of acid (to bring down a pH that is too high) or alkaline substances such as soda ash (if the pH is too low and needs to come up). With plaster spas, the calcium in the plaster creates a situation where the water is always in contact with alkaline material, always slowly dissolving it (remember, water is the universal solvent), causing the water to be slightly alkaline. Therefore, in plaster spas you will always be adding small amounts of acid to keep the pH down.

Acid is most readily available in liquid (muriatic acid) or dry granular forms. *Never* mix acid and chlorine—the result is *deadly* chlorine gas. Since each product is different, I won't try to guide your use of acid; just follow the instructions on the package or in your test kit for the amount to add. Keep this general guideline in mind, however: If you add acid in small amounts, you can always add a bit more to get the desired results. But if you add too much, the water becomes corrosive and so you must now add alkaline material to bring the pH back up.

Total Alkalinity

Total alkalinity will tell you the quantity of alkaline material (such as dissolved plaster) in the water. A pH reading is similar to a temperature reading on a thermometer, which simply states the present temperature, while a total alkalinity measurement represents the volume of heat that brought you to that temperature. In other words, total alkalinity is a measurement of the alkaline nature of the water itself and, therefore, the ability of that water to resist abrupt changes in pH.

The actual method of testing is discussed below. However, it is important to know at this point that an appropriate reading for your water's total alkalinity is 80 to 150 ppm. Below 80 ppm means that too much acid has been added even if the pH reading is high. Therefore, you always adjust first the total alkalinity level, then the pH. Proper maintenance of the total alkalinity of a spa will pay great dividends. You will use less chlorine and less acid and will suffer fewer algae problems.

Hardness

Hardness (or calcium hardness) is one component of the total alkalinity of the water. It is a measurement of the amount of one alkaline, soluble mineral (calcium) out of the many that may be present. In sufficient quantity in the water, calcium readily precipitates out of solution and forms deposits on the spa walls and tile and in equipment. This deposit is called *scale* and is the white discoloration you see at the waterline.

The acceptable range when you are testing for hardness is 100 to 600 ppm. Over 600 ppm and you will see the scale. The only "cure" for water that "hard" is to drain the spa partially or totally and to add fresh water.

Total Dissolved Solids

Now that we've talked about two measurements of minerals in the water, there is one overall category that helps you keep track of the "big picture." The total dissolved solids (TDS) are, as the name implies, a measurement of everything that has gone into the water and remained (not been filtered out), intentionally or not. The main contributor to TDS increases in a spa is simple evaporation. When the water evaporates, it leaves its contents behind. You add more water, so the solids keep building up as the liquid evaporates away.

In most places, water from the tap already has a reading of 400 ppm of total dissolved solids. You will add around 500 ppm more each year with chemicals. Depending on where you live, the evaporation rate may add another 500 ppm per year. Therefore, we say that when a body of water has reached a reading of 2000 to 2500 ppm, it's time to empty it. There is no other way to effectively remove all those solids floating around in your water.

Algae

Algae are one-celled plants, of which there are over 20,000 known varieties. Algae include microscopic ocean plankton, giant kelp that grows 2 feet per day, and virtually everything in between. Sunlight speeds algae growth, appearing in the spa as a green, brown, yellow, or black slime often resembling fur. It thrives in corners and on steps, where circulation may not be as thorough as elsewhere in the spa and so they are not being bombarded with chlorine or other sanitizers in sufficient strength to kill them. There are mostly three forms of algae that you will encounter in spas.

Green Algae (Chlorophyta) The most common algae is green and grows as a broad slime, which can be removed by brushing and proper sanitizer residual levels.

Yellow Algae (Phaeophyta) Yellow algae also at times appears brown or muddy in color and is therefore called *mustard algae.* It does not grow as rapidly as green algae but is more difficult to kill. It grows with the same broad, fur or moldlike pattern as green algae. Brushing has little visible effect, although it will remove the outer slimy layer, exposing the algae underneath to sanitizers.

Black Algae (Cyanophyta) Black algae is actually dark blue-green and is the pool technician's worst nightmare. It grows first in small dots, appearing to be specks of dirt. As time passes, these specks begin to enlarge and then appear all over the spa. Part of the reason for black algae's virulence is that instead of creating a slimy substance as a protective barrier as do green or yellow algae, black algae cover themselves with a hard substance that resists even vigorous brushing. Only a stainless steel brush will break open that shell, allowing sanitizers to penetrate the plant itself.

Algae Elimination Techniques

If there is any single reason for preventive maintenance on spas, algae growth is that reason. Maintaining proper water chemistry, balance, and a clean body of water are the best safeguards so that you will not need to deal with remedies instead.

Often the cause of an algal bloom can be attributed to trying to save money on the electric bill and shortening the circulation time of the system. Weather extremes or a heavy bather load can also add nutrients for algae and hasten decomposition of sanitizers.

There are two approaches to dealing with algae once they hit. One is a general elimination program that will work with most algal blooms. The other is to combine that program with a special algaecide.

General Algae Elimination Procedure

RATING: EASY

The following steps will help with any type or spread of algal growth and should be undertaken before you add algaecides.

1. **Clean the Spa** Dirt and leaves will absorb sanitizers, defeating the actions you are about to take. Be sure you have a clean skimmer and strainer basket and break down the filter as well, giving it a thorough cleaning. As the algae die or are brushed from the spa surfaces, they will clog the filter, so it is important to start clean. You may need to repeat this process several times during a major algae "cure."

2. **Check the pH** Adjust, if necessary (see "Water Testing" section below).

3. **Shock** Add 1 pound (373 grams) of trichlor for every 3000 gallons (11.4 cubic meters) of water. Brush the entire spa thoroughly, stirring up the trichlor for even dissolution and distribution. Never use trichlor or other granular sanitizers at such strength on dark-colored plaster or painted spas. These surfaces will discolor. Instead, use liquid chlorine, as outlined in the section on superchlorination.

4. **Circulate** Run the circulation for 72 hours, allowing the trichlor to attack the algae at full strength, brushing the spa at least once each day. Adjust the pH as needed, and continue to add liquid chlorine to maintain at least 6 ppm residual.

5. **Clean Up** As noted above, the dead algae may clog the filter, requiring teardown and cleaning several times. Keep an eye on the pressure. When the chlorine residual returns to 3 ppm, resume normal mainte-nance. You will need to vacuum the spa frequently during this period to remove the dead algae and inert ingredients of the trichlor, both of which will appear as white dust when you brush or otherwise disturb the bottom. You will also need to brush frequently, not only at first, but for at least 1 week after you can no longer see any trace of algae. Believe me, it's still there and will rebloom if you let up.

Special Algae Fighting Procedure

RATING: ADVANCED

The second general procedure to understand is for particularly stubborn algae, such as yellow (occasionally) or black (always). This method applies even when you can see only a few small patches, because there are other contaminated areas that you can't see yet.

1. **Prepare** Follow the routine outlined in the "General Algae Elimination Procedure" (above).

2. **Brush** Use a stainless steel wire brush, and vigorously brush the surface of the patches. Brush the remainder of the spa normally.

3. **Shock** In addition to the chemical application already recommended, pour some trichlor directly over the top of the algae spots. You may need to shut off the pump to accomplish this, so that the currents in the spa don't redistribute the trichlor you are trying to place on top of the algae. If the algae is on a vertical location, fill a stocking with trichlor or tablets and hang it on the spa wall so that the stocking makes contact with the spots. It isn't necessary to cover every inch of every algae location. The act of locating such a concentrated dose of chemical within a few inches will have a similar effect.

4. **Add Algaecides** Application of each product will differ, so follow the package instructions. The various types of algaecides, based on their active ingredient(s), are as follows:

   - **Copper Sulfate (Liquid Copper)** Effective on all types of algae, but recommended for use only in dark-colored spas due to potential staining.

   - **Colloidal (Suspended) Silver** Effective against all types of algae in all situations.

   - **Polymers** Effective on all types of algae to some degree, but best on green and yellow algae.

   - **Quats (Quaternary Ammonium Compounds)** Good on green algae in early stages.

5. **Keep Brushing!** Continue daily brushing of the spot with the wire brush, and reapply chemicals until the algae are gone.

TRICKS OF THE TRADE: COMMON SPA WATER PROBLEMS AND CURES

Cloudy water

- Inadequate filtration
  - Increase daily filtration time.
  - Clean the filter.
- Too many total dissolved solids
  - Check the total alkalinity and pH; adjust as needed.
  - Drain and refill with clean water.

Algae in spa

- Follow "Algae Elimination Techniques" as described.

Blue-green water

- Copper from plumbing or heater components, stripped away by acidic water
  - Immediately shut down the circulation equipment until you can ascertain the problem and solution.
  - Test and balance pH. Apply alkaline (pH raisers) and brush the spa. After several hours, take another pH reading. If it is now normal, turn the equipment back on; otherwise, keep balancing.
  - Apply a metal chelation agent to the water. These products, available in various brand names, will attract and combine metals so the metals can be filtered out. Follow label instructions of the product you choose, and in all cases, run the filter for 72 hours once you have balanced the chemistry.

Brown-red water

- Iron in the water from metal fixtures corroded by acidic water
  - Check and adjust first the total alkalinity, then the hardness, and finally the pH.
  - You may need to drain part of or all the water and add fresh.
  - Apply a metal chelation agent to the water (see above).

Corrosion of metal light fixtures, rails, etc.

- Water too acidic: Take steps outlined above
- If metal turns black, electrolysis may be the problem.
  - Look for electric current (perhaps a slight water leak in a light fixture or J-box); and water with enough minerals (salts) to conduct the weak current.

- Correct as needed.

Scale forming on spa walls or at waterline

- Buildup of calcium carbonate precipitated out of water from evaporation or heat
  - Check the hardness. If it is near or exceeds the standard of 2000 ppm, drain and replace some of or all the water.
  - If the hardness is within acceptable limits, the problem may be high pH or total alkalinity. Check and adjust both.

Eye/skin irritation; colored hair or skin

- Low pH and/or too many chloramines: Adjust the pH and shock-treat the spa

Odors

- Chlorine odor from too many chloramines: Adjust the pH and shock-treat the spa
- Musty odor from algae growth or high bacteria: Shock-treat the water to kill any bacteria or drain and refill spa
- Mildew odor from accumulation of mildew on spa covers or in deck crevices where water has been standing: Follow your nose to the source and clean contaminated area with sanitizers

Foamy water

- Too much soap, body oil, or lotion from bathers; too much of spa cleanser
  - Extend filtration time until foam is gone; clean filter.
  - Drain and refill spa.

Water Testing

Many manufacturers produce spa test kits with exact directions. In the typical multitest combination pool/spa test kit, you will find five small squeeze bottles, a color chart, instructions, and a sample beaker (Fig. 7-2).

Testing Methods

There are three basic approaches to water testing: using chemical reactions and comparing the resulting colors; estimating values with electronic devices; or making observations of the relative cloudiness (turbidity) of a water sample.

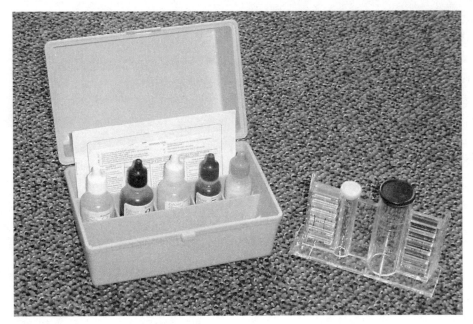

FIGURE 7-2 Typical spa water test kit.

COLORIMETRIC

The most common approach to testing water is to collect a small sample in a clean tube and add some chemicals to the sample. These chemicals, called *reagents,* will evaluate the sample by its changing color. By comparing the intensity of the color with that on a color chart of known values, we can determine the relative degree of each water chemistry parameter. For example, in one such test, a sample turning yellow denotes the presence of chlorine. The stronger the intensity of the color, the more chlorine is in the sample.

TITRATION

Also a color-based test, *titration* is the process of adding an indicator reagent (a dye), followed by a second reagent (called the *titrant*) in measured amounts, usually a drop at a time, until a predicted color change occurs. By noting the number of drops of the reagent required to effect the color change, we can estimate the value in question. Some manufacturers supply tablets in place of liquids for titration testing. For example, the amount of acid demanded by a body of water can be calculated with this method.

TEST STRIPS

Paper strips impregnated with certain test chemicals (Fig. 7-3) are also used as colorimetric test media. With this method, you simply dip a test strip directly into the spa and move it around in the water for 30 seconds. When you remove the strip, you compare its color to the color chart of known values to determine the test results. Test strips are produced for individual tests or for several on one strip. Test strips are more costly to use, but are popular because of their simplicity and accuracy.

ELECTROMETRIC

Testing using electronic probes, attached to calibrated digital or analog readouts, is the most accurate method of chemical testing. Figure 7-4 shows a unit that can be simply dipped in the water being tested.

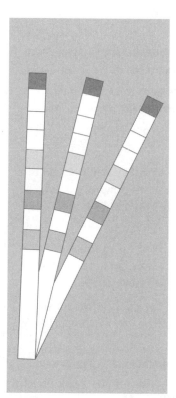

FIGURE 7-3 Test strips.

FIGURE 7-4 Electronic water testing unit.

TRICKS OF THE TRADE: ACCURATE WATER TESTING

RATING: EASY

- Use only fresh reagents or test strips. Keep them out of prolonged direct sunlight and store at moderate temperatures. Look at reagents before use. If they have changed normal color, appear cloudy, or have precipitate on the bottom of the bottle, discard them.

- Rinse the sample vials and any other testing equipment several times thoroughly, using the water you are about to test. Never clean equipment with detergents as chemical residue from these products can deliver false readings.

- Consider where and when you are sampling to ensure truly representative results. Samples should be collected several hours after any chemicals have been added and thoroughly circulated through the body of water, to ensure that their effect has been completely registered. In any case, the water should have been circulating for at least 15 minutes before sampling. Sample away from return outlets to make sure you are collecting a representative sample.

- Collect samples from at least 18 inches (45 centimeters) below the surface to avoid inaccuracies caused by water evaporation, direct sunlight contact, dirt, etc.

- Make color comparisons in bright, white light to ensure accurate colorimetric comparisons, against simple white backgrounds. Most test kits come with a white card to place against the back of the test vial.

- Observe and record results immediately for accurate results.

- Handle samples and reagents carefully. Some reagents can cause skin irritation, so avoid direct contact and never pour them into the spa, even after the test is complete. Avoid contact with the sample, since acids and oils from your hands can contaminate a sample, leading to false readings.

- Take your time. Hurried testing will lead to inaccurate results, leading to improper application of chemicals. Follow the directions in the kit. Some may seem redundant or unnecessary, but, believe me, they are stated for a reason, so follow them.

- Never *flash-test*—dashing a couple of drops of each reagent into the spa. There is a very different color appearance when you view a dispersing reagent in the spa as compared with a captured result in a vial held at eye level.

- Conduct all tests before you add any chemicals to modify the water chemistry. For example, if you test for chlorine residual, then add significant amounts of chlorine in any form, and finally test the pH, then your pH reading will reflect the value of the chemicals just added, not that of the water itself. After adding chemicals to the water, allow adequate time for distribution and test again to see if your actions were adequate. Allow at least 15 minutes for liquid chlorine to circulate and at least 12 hours for the pH to adjust before testing again.

- Since test kit reagents can be replaced, you will probably use your kit for many years. The color chart, however, may fade over time. Compare your chart against a new one from time to time, or simply buy a new test kit annually (they're not that expensive).

Testing for Each Water Quality Parameter

Let's examine each parameter of water quality and identify which test is used for each.

SANITIZERS

Several reagents are used for testing for the presence of not only chlorine, but also the free, available chlorine within the total. The first bottle in most reagent test kits is OTO (orthotolidine). When combined with a sample that contains chlorine, OTO will turn color, from light pink to yellow to deep red, depending on the strength of the chlorine. When comparing that color of the sample with the color chart, we can determine the exact concentration of chlorine in the water (expressed in parts per million).

The only drawback to this simple procedure is that OTO measures only total chlorine content. It does not distinguish between free, available chlorine and that which is present but combined with other contents to form chloramines. It is, therefore, a good "indicator" test, but not a complete result. OTO can also measure bromine in the same way, using a slightly different color chart and with similar limitations. Bromine results will turn a darker yellow.

A second colorimetric test, which determines the actual free chlorine in the sample, is conducted with a reagent called DPD (diethyl phenylene diamene). By subtracting the results of the DPD test from those of the OTO test, you will know the amount of combined chlorine, chloramines, in the sample as well.

DPD produces colors between pink and red for both chlorine and bromine. DPD reagents are produced as liquid or tablets, with several manufacturers providing additional reagents for determining the results of total, combined chlorine and then singling out free, available chlorine (or bromine). Each manufacturer provides detailed, simple instructions for its particular products.

PH

Reagents and a color comparison chart are also used for testing pH. The reagent for this test is phenol red (phenosulfonephthalein). Before you add phenol red to a sample, however, any chlorine in the sample must be neutralized to avoid a false reading, using a predetermined amount of the neutralizer sodium thiosulfate, which is provided in your test kit. All test reagents used should be fresh, to ensure accurate results; however, phenol red is the one that will quickly go bad with age.

Perhaps the best use of electronic testing devices is in determining the pH. They are accurate, compensate for extreme temperatures (useful when testing spa water), and are not subject to problems from high chlorine or bromine levels.

TOTAL ALKALINITY

The most common test is the titration method described above. There are no significant limitations of total alkalinity titration tests such as those described for other reagent tests. Test strips are also available for total alkalinity testing.

HARDNESS

Hardness is tested using colorimetric and titration tests combined or test strips. With combined testing, a buffer solution is added to the sample to increase the pH to around 10.0 to facilitate the accuracy of the subsequently added dye solution. The dye reacts with calcium and magnesium in the sample to produce a red sample. A reagent called EDTA (ethylenediamine tetra-acetic acid) is then added a drop at a time until the solution turns blue. The number of drops is compared to a chart to evaluate the hardness of the sample.

TDS

There is no simple, inexpensive way to test for total dissolved solids in a water sample, so take a water sample to a local spa retailer that is equipped with electronic equipment to analyze for TDS. With the recent advent of less costly electronics, however, TDS meters have dropped below the $100 mark and may be worth the investment, if combined with a pH meter and thermometer probe.

Application of Chemicals to Spa Water

There are as many products for sanitizing and balancing the chemistry of water as there are for testing it. It pays to follow the directions on product labels, but some general guidelines apply.

Liquids

RATING: EASY

Liquid chlorine and muriatic acid are generally sold in gallon plastic bottles. When adding liquid chlorine or acid to a spa, pour as close as possible to the water's surface to avoid splashing it on your shoes or the deck, discoloring both. Air and sun contact will diminish the concentration of chlorine, so keeping it close to the surface of the water also minimizes those impacts. Add the chlorine slowly near a return line while the circulation is running, to maximize even distribution. Never pour liquid chlorine into the skimmer. Avoid skin or eye contact with chemicals.

Granular Products

RATING: EASY

Granular products tend to settle to the bottom of a spa before dissolving, so if the vessel is dark plaster, it may be wise to brush immediately after application. None of these products should be poured into the skimmer directly, and all should be applied with the circulation running. Since granular products tend to be extremely concentrated, they can cause skin irritation or breathing problems if direct contact is made. Handle them with care. Granular products tend to have longer shelf lives than liquids; however, time or prolonged exposure to sunlight will diminish their efficacy as well. Store and treat granular product as you would liquid.

Tabs and Floaters

RATING: EASY

Chlorine tablets are sold to place in floating devices (sometimes shaped like ducks) which allow a slow dissolving process. Tablets are valuable when you can't service the spa for some time and need a constant source of sanitizer. Some floaters have a valve that allows more or less water to flow in them, theoretically controlling the amount of chemical entering the water, but the results are trial and error. Like granular products, tablets left on the bottom of a spa will bleach out any color, so use a floater unless the surface is already white. Never leave a tablet in the skimmer. The low pH of tablets means you are assaulting your circulation equipment and any related metal plumbing with acid, and we have already discussed the perils of that.

Mechanical Delivery Devices

RATING: ADVANCED

Chlorine is the only spa chemical generally added by mechanical device, although large commercial pools may use such devices for other chemicals as well. Mechanical delivery systems generally fall into two categories, erosion systems or pumps. Pumps are generally found on very large spas or commercial units, but simple erosion systems can be installed and maintained by anyone.

As the name implies, the erosion system uses the water passing over the dry chemical to erode or dissolve it into solution and thereby into the receiving water. Like the floater, the erosion system is usually controlled by restricting the volume of water allowed to pass through the system and, therefore, the amount of erosion that can take place of the tablet or granular chlorine (or bromine) inside. Figure 7-5 shows a typical erosion chlorinator system. Chlorine tablets are placed in the drum, and the water passing over dissolves them.

Erosion systems are typically made of PVC plastic and are located in the equipment area, plumbed directly in the circulation lines. These chlorinators are easily installed by following the directions supplied, requiring only basic plumbing techniques. They must always be located after any other equipment in the system. If you were to place them, say, between the filter and heater, the concentrated chemical would corrode the internal parts of the heater.

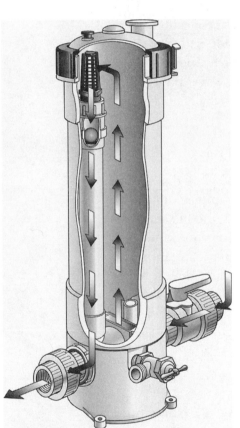

FIGURE 7-5   In-line erosion-type chlorinator.
Pentair Pool Products, Inc.

Cleaning and Servicing

Now that we have reviewed what makes a spa "tick," including the components of water chemistry itself, we turn to the most basic aspect of water maintenance—routine cleaning and servicing. We first review the tools of the trade and then describe how to use them in a typical service procedure call and in special situations.

Tools of the Trade

Figure 8-1 shows the basic service equipment carried by a professional water technician and the do-it-yourself spa owner alike. Let's examine the utility of each item.

Telepole

Made of aluminum or fiberglass, the telepole (telescoping pole) is the backbone of the spa cleaning system. The one used most for spas is 4 feet (1.2 meters) long, telescoping to 8 feet (2.4 meters) by withdrawing the inner pole out of the outer one (Fig. 8-2). The two poles are locked together by twisting them in opposite directions, engaging a cam lock or compression ring nut. The end of the pole will have a handgrip or a rounded tip to prevent the hand from slipping off the pole.

At the end of the outer pole you will notice two small holes drilled through each side of the end, about 2 inches (5 centimeters) from the

FIGURE 8-1 Typical cleaning and servicing tools.

end and again about 6 inches (15 centimeters) higher. The diameter of the pole is designed to fit the various tools you will use; you attach them to the pole by sliding the end of the tool into the end of the pole, where it is held in place with small clips. Other tools are designed to slip over the circumference of the pole, but they also use a clip device to secure the tool to the holes at the end of the telepole.

Leaf Rake

Figure 8-3A shows a professional, deep-net leaf rake. The net itself is made from stainless steel mesh, and the frame is aluminum with a generous 16-inch-wide (40-centimeter) opening. The leaf rake shank fits into the telepole and clips in place as previously described. Be careful not to spill acid or other caustic chemicals on your leaf rake; either the metal or plastic mesh will deteriorate and holes will develop. Some leaf rakes are designed to disassemble and replace the netting as needed. Figure 8-3B shows a typical shallow surface skimmer, designed for lighter duty.

Wall Brush

The wall brush (Fig. 8-3C) is designed to brush the interior surfaces of the spa. It is made of an aluminum frame with a shank that fits the telepole, and the nylon bristles are built on the brush either straight across or curved slightly at each end. The curved unit is useful for getting into corners and tight step areas. Wall brushes come in various sizes, the most common for spa use being 6 to 10 inches (15 to 25 centimeters) wide. Brushes are also made with stainless steel bristles for heavy stains or algae problems. Always use a stainless steel brush if you need the added strength of metal. Common steel bristles can snap off during brushing and leave stains on the plaster or plastic when they rust.

Vacuum Head and Hose

There are two ways to vacuum the bottom of a spa. One actually sucks dirt from the water and sends it to the filter. The other uses water pressure from a garden hose to force debris into a bag which you then remove and clean (see "Spa Vacuum" below).

The vacuum head and hose (Fig. 8-4A and B) are designed to operate with the spa circulation equipment. The hose is attached

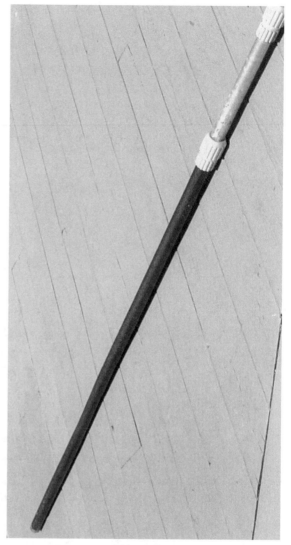

FIGURE 8-2 Telepole.

at one end to the skimmer suction port and at the other end to the vacuum head. The vacuum head is also attached to the telepole. With the pump running, you can now glide the vacuum head over the underwater surfaces, vacuuming up the dirt directly to the filter.

Vacuum heads are made of flexible plastic, with plastic wheels or a brush that keeps the head just above the spa surface. The flexibility of the head allows it to contour to the curvature of spa corners and bottoms.

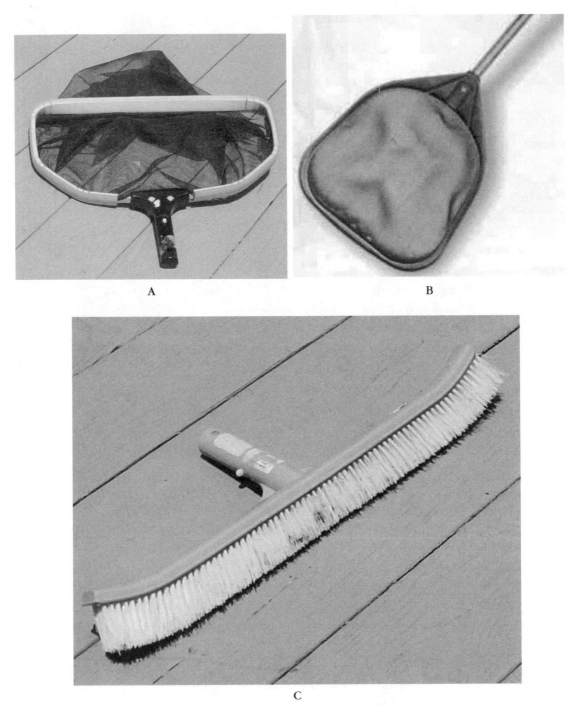

A

B

C

FIGURE 8-3 (A) Deep leaf rake. (B) Surface skimmer. (C) Wall brush.

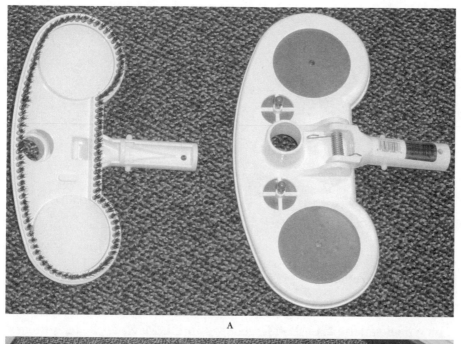

A

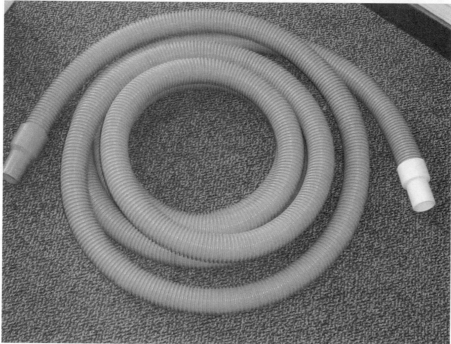

B

FIGURE 8-4 (A) Vacuum head. (B) Vacuum hose. (C) Spa vacuum and two-part telepole.

C

FIGURE 8-4 *(Continued)*

The wheel version is best on larger spas and those with plaster surfaces; the brush model is best on plastic surfaces that might be scratched by wheels. Figure 8-4C shows a vacuum head and two-part telepole designed especially for spa work, which connects to the standard vacuum hose.

Hoses are also available in economy models (thin plastic material) through "Cadillacs" (heavy rubber plastic material with ribs to protect against wear) and in various lengths (10 to 50 feet, or 3 to 15 meters). The hose cuff is made 1¼-inch (32-millimeter) or 1½-inch (40-millimeter) diameter, to coordinate with similar vacuum head dimensions. Cuffs are female-threaded at the end that attaches to the hose itself, so you can screw replacements onto a hose. The best cuffs allow them to swivel on the end of the hose so that when you are vacuuming, there is less tendency to coil and kink the hose.

Spa Vacuum

Figure 8-5A shows a spa vacuum (called a *spa vac*), that attaches to the telepole and a garden hose and operates by forcing water from the hose into the unit, where it is diverted into dozens of tiny jets, which are directed up toward a fabric bag on top of the unit. The upwelling water creates a vacuum at the base, sucking leaves and debris into the unit and up to the bag. Water passes through the mesh of the bag, but the debris is trapped.

Fine dirt will pass through the filter bag; however, a fine-mesh bag is sold for these units, and simply double-bagging will capture more dirt. I use a fine-mesh sock or ladies' hosiery, attached to the spa vac inside the bag provided for finer filtering of dirt.

There are two other types of spa vac. The hand-pump model creates its own suction (Fig. 8-5B), and a battery-powered unit has a built-in pump.

Tile Brush and Tile Soap

Tile brushes are made to snap into your telepole so you can scrub the tile without too much bending. Mounted to a simple L-shaped, two-part aluminum tube, the brush itself is about 3 inches × 5 inches (8 centimeters × 13 centimeters) with a fairly abrasive foam pad for effective scrubbing.

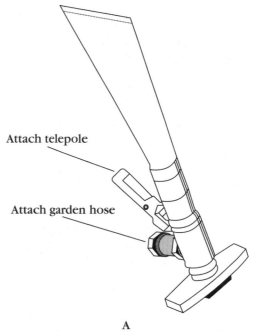

Attach telepole

Attach garden hose

A

FIGURE 8-5 (A) Spa vacuum. (B) Surface skimmer and spa vacuum.

I have found that these brushes are valuable for wiping a bit of algae off ladders or other tricky spots in spas, but the elbow grease required to remove body oil, suntan oil, and scale from tiles is much more than you can get at the end of the telepole. Therefore, I also carry a barbecue grill cleaning pad. It has a convenient grip handle and an abrasive Brillo-type pad, much more effective at cleaning tiles. Since this will also require getting on your hands and knees around the entire circumference of the spa, carry a foam kneepad as well!

Tile soap is sold in standard preparation at the supply house, but I recommend mixing it into another container with 1 part muriatic acid for every 5 parts soap. This will help cut the stubborn stains and oils, but it will also eat into the plastic on the tile brush pads and the plastic of the barbecue brush handle, so keep rinsing them in spa water after each application and scrubbing. Don't use other types of soap in place of tile formulations, since they may foam and suds up when they enter the circulation system (especially in spas).

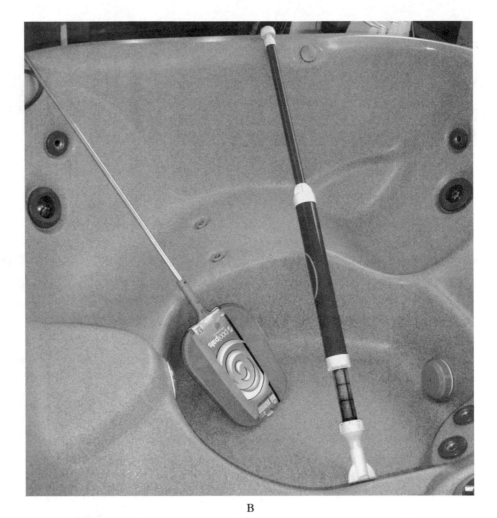

B

FIGURE 8-5   *(Continued)*

Test Kit

As discussed in Chap. 7, buy the best test kit you can afford and keep it in good working order. Your chemical testing is by far the most important aspect of maintaining the spa. Since test kits and methods were discussed previously, I will not repeat the information here.

Pumice Stones

The soft pumice stone, made from volcanic ash, is abrasive enough to remove scale from tiles and other deposits or stains from plaster surfaces without scratching them excessively. Pumice stones are sold as blocks, about the size of a brick, and as small "bladed" stones which attach to your telepole for reaching tight spaces and underwater depths. Since pumice stones disintegrate easily, it is wise to use them before you vacuum a spa. Alternately, you can brush the residue to the main drain where it will be carried to the filter.

Spa Cleaning Procedure

RATING: EASY

Every technician, and every homeowner for that matter, will approach spa cleaning differently. Yet after many years and hundreds of thousands of service calls, I have discovered that there are really a few basic procedures that are efficient and save time which any technician would do well to follow. I present my procedures and reasons below.

1. **Clean the Deck and Cover** A quick sweep or hosing of any debris near the spa (Fig. 8-6A), at least 10 feet back from the edge, will keep your service work looking good after you have left. Similarly, remove as much debris as possible from the spa cover before removing it. If the cover is a floating type, be sure to fold or place it on a clean surface; otherwise, when you put it back in place, it will drag leaves, grass, or dirt into the spa.

2. **Adjust the Water Level** Check the level of the spa with the water calm (Fig. 8-6B). If several inches of water have evaporated since your last servicing, the spa may take some time to refill. Therefore, it's a good idea to put the hose into the spa and start the process immediately.

A

B

FIGURE 8-6 (A) Hosing the deck. (B) Checking the water level. (C) Cleaning the spa. (D) Scrubbing the waterline (or tiles). (E) Vacuuming with a spa vac.

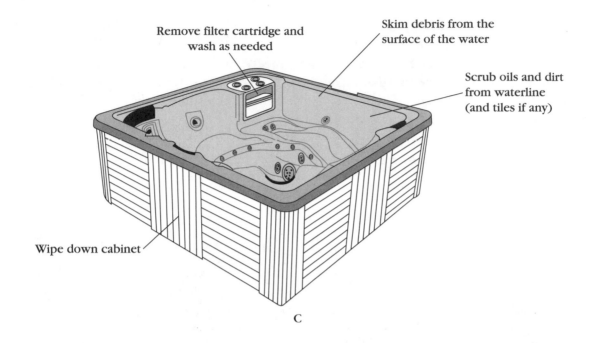

Remove filter cartridge and wash as needed

Skim debris from the surface of the water

Scrub oils and dirt from waterline (and tiles if any)

Wipe down cabinet

C

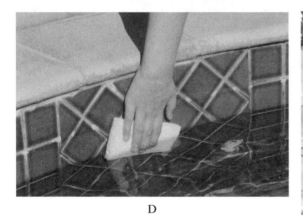

D

E

FIGURE 8-6 (*Continued*)

3. **Skim the Surface** It is much easier to remove dirt floating on the surface of the water than to remove it by any means from the bottom. Using your leaf rake and telepole, work your way around the spa, "raking" any floating debris off the surface (Fig. 8-6C). As the net fills, empty it into a trash can or plastic garbage bag. Never empty your skimming debris into the garden or on the lawn—the debris is likely to blow right back into the spa as soon as it dries out.

   As you skim, scrape the tile line, which always acts as a "magnet" for small bits of leaves and dirt. The rubber-plastic edge gasket on the professional leaf rake will prevent scratching of the tile. This action is often overlooked and is a rich source of small debris that will soon end up on the bottom. If there is a scum or general dirt on the surface, squirt a quick shot of tile soap over the width of the spa. The soap will spread the scum toward the edges of the spa, making it more concentrated and therefore easier to skim off.

4. **Clean the Tiles** Many technicians leave the tiles for last, but if they're fairly dirty, you will remove the material from the tile and it will settle on the bottom that you have already cleaned. Also, if you need to remove stubborn stains with a pumice stone, the pumice itself breaks down as you scrub, again depositing debris on the bottom. Therefore, clean the tiles (or the waterline) first (Fig. 8-6D). Generally, even if the tiles appear clean, I scrub them once a month anyway to knock loose any scale or other deposits before they become noticeable. Whichever brush you use, apply a squirt of tile soap directly to the brush and start scrubbing. Remember, don't use soaps not designed for tile work as they will foam up. I carry a foam kneepad, available at any gardening store, and work my way around the spa. Scrub below the waterline as well as above (Fig. 8-6D). Evaporation and refilling mean the waterline is rarely always at the same level, so clean the entire tile line.

5. **Check the Equipment** Review the circulation system by following the path of the water.

   ■ Since you have already skimmed and cleaned the tile, you can now clean out the spa's skimmer basket without concern that it will fill up again while you work.

■ Next, open the pump strainer basket and clean that. If the pump has a clear lid, you may be able to observe whether this step is necessary. Reprime the pump.

■ Check the pressure on the filter. If the pressure is high, the filter may need cleaning.

■ Look over the heater for leaves or debris. Look inside to make sure rodents haven't nested and to verify the pilot light is still operating (if the unit is a standing pilot type). Turn the heater on and off a few times to make sure it is operating properly.

■ Check the time clock. Is the time of day correct? Is the setting for the filter run long enough for prevailing conditions? Always check the clocks, because trippers come loose and power fluctuations or outages can play havoc with them.

■ At each step of the equipment check, look for leaks or other early signs of equipment failure.

■ Clean up the equipment area itself. Remove leaves from around the motor vents and heater to prevent fires; clear deck drains of debris that could prevent water from draining away from the equipment during rain.

6. **Vacuum the Spa** If the spa is not dirty or has only a light "dusting" of dirt, you may be able to brush the walls and bottom, skipping the vacuuming completely. Otherwise you have two ways to clean it: vacuuming to the filter or vacuuming with the spa vac.

A. **Vacuum to the Filter** As the term implies, *vacuuming to the filter* means the dirt is collected from the spa and sent to the circulation system's filter.

■ Make sure the circulation system is running correctly. If the system includes valves for diversion of suction between the main drain and skimmer, completely close the main drain valve and turn the skimmer valve completely open.

■ Attach your vacuum head to the telepole, and attach the vacuum hose to the vacuum head. Working near the skimmer, feed the head straight down into the spa with the hose following. By slowly feeding the hose straight down, water will fill the hose and displace the air. When you have fed all the hose into the spa, you should see water at the other end (which should now be in your hand).

■ Keeping the hose at or near water level to avoid draining the water from it, slide the hose through the skimmer opening and into the skimmer. Attach the hose to the skimmer's suction port. The hose and vacuum head now have suction.

■ Vacuuming a spa is no different from vacuuming your carpet. Work your way around the bottom and sides of the spa, but avoid moving the vacuum head too quickly. You will stir up the dirt rather than suck it into the vacuum.

■ When finished, position yourself near the skimmer and remove the vacuum head from the water. The suction will pull the water rapidly from the hose. Be prepared to pull the hose off of the skimmer suction port before the air reaches there; otherwise, you will lose prime. You can also let the vacuum run for a minute after you finish to make sure the hose contains clean water; then remove the hose from the skimmer suction port and drain the water from the hose back into the spa.

■ Remove your equipment from the spa. Check the pump strainer basket and filter to see if they have become clogged with debris from vacuuming. Clean as needed. Replace the skimmer basket.

B. **Vacuum with the Spa Vac (Fig. 8-6E)** If the spa is littered with leaves or other heavy debris, you may need to use the spa vac instead (or the spa vac for heavy debris, followed by the vacuum to filter to capture the finer dirt).

■ Attach the garden hose to a water supply, then to the spa vac. Clip the spa vac onto the telepole. Always be sure your collection bag is securely tied to the spa vac itself. I can't tell you how frustrating it is to have the bag come loose as you lift the spa vac from the spa, spilling the debris back into the area you just cleaned.

■ Insert the spa vac into the spa. Turn on the water supply and vacuum. As with vacuuming to the filter, spa vacuuming is just a matter of covering the spa floor and walls. Be careful not to move the spa vac so fast that you stir up the debris

TRICKS OF THE TRADE: ENVIRONMENTALLY FRIENDLY SPA MAINTENANCE

RATING: REALLY EASY!

Spas use fossil fuels to heat them and toxic chemicals for sanitation and cleaning, and spas consume vast amounts of water. So can we enjoy the many benefits of a spa with a clear environmental conscience? Yes, if we follow a few basic guidelines for conservation and proper waste disposal—reduce, reuse, and recycle.

Chemicals

- Chlorine and other chemicals should be purchased in amounts likely to be used within a week (or two at most) so that you are more likely to use the entire amount.
- Buy products in reusable/recyclable containers.
- Use nontoxic, biodegradable alternatives to harsh cleansers.

Energy conservation

- Install solar heating.
- Double-cover your spa to trap more heat and reduce evaporation.
- Create windbreaks around the spa to prevent heat loss and evaporation with fencing, shrubs, walls, or rockscapes.
- Insulate the exterior of the spa (set it in-ground if it isn't already).
- Don't leave the light on except when you're using the spa.
- Shut off the heater completely if you're not likely to use the spa for several days.
- Insulate pipes.
- Don't run pumps longer than necessary. The average spa pump uses as much energy each hour as the average window air conditioner.

Water conservation

- Avoid excessive splashing. Still water evaporates at one-half (or less) the rate of disturbed water.
- Use an unheated spa instead of your home air conditioner for relief on hot days. Take a dip to cool yourself instead of cooling the entire house.
- Check frequently for leaks. You can waste a lot of water before realizing you have a problem unless you keep a good watch on the equipment area and the water level.
- Don't hose off decks. Use a broom.
- Direct rainwater to the spa for refilling.

Recycle

- Save the leaves taken from your spa for mulch.

- Buy equipment and supplies with the greatest amount of postconsumer recycled materials. Even plastic pumps and filters are now being made with recycled materials!

- Never throw away an old piece of equipment. Take the motor to a rebuilder (who can use the parts and may even give you a few dollars for it); take the old filter or heater to a scrap metal dealer.

- Replace the net on your skimmer instead of buying a new one. Other maintenance tools may have replacement components, so try before you buy.

- Use discarded stockings or panty hose as catch bags on spa vacs.

Miscellaneous

- When painting a spa or deck, use paints with low volatile organic compounds (VOCs) to protect air quality.

- When sanding, chipping, or grinding, control the dust and properly dispose of all residue. It may contain fine particles of old paint, metals, or chemicals which can pollute the air you are breathing in the vicinity or harm plants and pets.

- When landscaping around the spa, avoid trees that drop lots of leaves; and don't create dirt planter areas near the water where wind and rain will flush debris into the spa. Organic debris causes more chemical use, and as noted above, you're trying to cut down.

- In some places (such as California), storm drains flow directly to rivers and the ocean. Thus anything that drains from your deck into the storm drains also flows to natural water courses. So don't use detergents or harsh chemicals to clean your spa deck unless you want to be swimming with the same pollutants!

ahead. If your garden hose is not the floating variety, work in a pattern that keeps the hose behind your work, so you will not stir up debris before you can vacuum it.

■ When finished, remove the spa vac by turning slightly to one side and slowly lifting it through the water to the surface. If you pull it straight up, debris will be forced out of the bag and back into the spa. Never turn off the water supply before removing the spa vac from the spa for the same reason. The loss of vacuum action will "dump" the collected debris right back into the spa. When the spa vac is on the deck, shut off the water supply and clean out the collection bag.

7. **Chemical Testing and Application** Follow the general testing guidelines described previously and apply the chemicals accordingly; then move ahead to the brushing step to give the chemicals time to circulate and distribute. The brushing will accelerate that distribution. Then apply acid or alkaline materials. You don't want to apply them together, since combinations of spa chemicals can be deadly.

8. **Brush the Spa** On spas that are not very dirty, you can skip vacuuming and brush from the walls to the bottom, directing the dirt toward the main drain, where it will be sucked to the filter. If you plan to use this technique instead of vacuuming, divert all suction to the main drain as previously described.

9. **Cleanup/Closeup** After you have collected all your belongings, take one last look at the spa before leaving. Did you add enough water and turn off the water supply? Did you pick up everything you brought with you?

Winterizing

If you work in a cold climate, I probably don't need to tell you that a body of water needs to be prepared for the winter months (Fig. 8-7). But even in warmer climates, you might discover that some of the follow-

FIGURE 8-7 A spa in winter.

ing information is useful. Water expands when it freezes, meaning that if it is trapped inside pipes and equipment, it will expand and crack them. PVC pipes, fittings, and ABS plastic pump parts are most susceptible, but soft copper heat exchangers and galvanized fill-line pipes are not exempt either. Expanding frozen water will also crack tiles and plastic skimmers.

The second problem, which relates to a winter seasonal closure of a spa, is potential damage from algae and debris. Although algae do not grow well in temperatures below 55°F (13°C), especially if the water is shielded from sunlight by a cover, prolonged periods of stagnation will permit and promote algae development.

Above-ground plastic spas can be drained completely, although inground gunite and large plastic spas should be treated just as pools are—the plumbing and equipment must be prepared for winter as described below.

Winterizing the Spa

RATING: MODERATE

1. **Balance the Chemistry** Etching or scaling conditions of water will harm the spa even when the circulation is off; so before closing it, make sure the chemistry of the water is balanced.

2. **Cleaning** Dirt and debris left in the water during long periods of stagnation will leave stains and/or be much harder to remove several months later. Therefore, thoroughly service and clean the spa before closure.

3. **Algae and Stain Prevention** To prevent algae growth and staining during the closure, superchlorinate the water. A simple formula is to triple the superchlorination that you usually use for the particular installation. A more precise approach is to raise the residual to 30 ppm for plaster surfaces and 10 ppm for vinyl or plastic. Add a metal chelating agent to prevent metals from dropping out of solution and staining the surfaces. Finally, add an algaecide that will inhibit black algae growth (I prefer silver-based products).

   Do not leave tablets or floaters in a spa during closure. Since the water isn't circulating, the chemical isn't dissolving either. The extreme concentrations in one area can do structural or cosmetic damage, particularly to vinyl or plastic surfaces.

4. **Shut Down the Equipment** Turn off the circulation equipment at the breakers, and tape them over to prevent someone's turning them back on. Turn off all manual switches and time clocks, and remove the trippers as an added precaution in case someone does turn the breakers back on. Pay special attention to lights, which may be on household circuits rather than wired through the equipment breakers. Some technicians don't disconnect underwater lights if the water level will remain above the fixture all winter. The light provides a little warmth to the water periodically and draws the

owner's attention to the spa for regular inspection. I have seen light lenses crack, however, as the extremely cold water and extremely hot light fixtures repeatedly contact, so I don't recommend this practice.

5. **Lower the Water Level** Pump out 24 to 36 inches (61 to 91 centimeters) of water, or at least enough to drop the level 18 inches (45 centimeters) below the tile and skimmer line. When water freezes and expands, it can crack the tile and plastic skimmer components, so the goal is to lower the level to a point where winter rain or snow will not raise it back up to those delicate areas. The spa should not be drained completely for the season, since hydrostatic pressures may cause cracks of the entire vessel to pop out of the ground.

6. **Clear the Lines** Perhaps the most important objective of winterizing is to protect the plumbing from freeze damage. There are several methods, some easier than others. The method you choose may depend on your equipment, skills, and the availability of water in your area.

   The most effective method of protection is to entirely drain the spa, making complete evacuation of the lines possible. Of course you would then refill it to the level discussed in step 5. If the water needs draining anyway for chemistry reasons, as discussed in Chap. 7, this is an opportunity to accomplish two tasks at once. The cost or availability of water in your area may prohibit annual draining; but if you can, it is the most effective way to be sure all water has been removed from the system. Even small amounts of frozen water can crack pipes and fittings.

   With the spa empty, remove the collar/nozzle fittings from return lines and plug them with rubber expandable test plugs. The main drain may still have water in the bottom of the plumbing, so try to suck it out with a wet/dry shop vacuum, or mop it out with a sponge. Just to be sure, pour a cup of antifreeze into the main drain before plugging it. For antifreeze, use a mixture of 1 part propylene glycol to 2 parts water. Your supply house will have propylene glycol or premixed products. Never use automotive antifreeze, which is corrosive. With all the lines plugged, refill the spa to the level described above.

   The return lines can also be filled with antifreeze by starting the pump with the strainer pot lid open. With the pump running, pour the mixture into the pump. It will be distributed throughout the equipment, plumbing, and return lines. When you can see antifreeze discharging from all the return lines, you can turn off the system and plug the return outlets. You may want to add some food dye to the mix to make it easier to see the concentration discharging to the spa.

7. **Remove Equipment** Any equipment in the spa or circulation system that may be damaged by exposure to the elements (or which may be stolen) should be removed and stored. The circulation equipment should be disassembled and important components stored. If the pump/motor is plumbed with unions, you can easily remove the entire unit and disconnect the electrical connection. If there are no unions, you might want to cut the plumbing and add the unions when you reinstall, to make the process easier next season. If the plumbing makes removal of the pump difficult, unbolt the wet end and motor from the volute and remove those.

Disassemble the filter, clean and thoroughly drain the tank, and put the grids or cartridges in storage. Freezing water can cause fabric deterioration. Sand filters should be cleaned and drained. Since rain might refill an open filter tank, close the tank and leave the drain plug out.

Shut the gas valve to the heater and any supply valve (if it is a dedicated line for the spa heater) at the meter. The heater has drain plugs on both sides. I recommend removing the heat exchanger and burner tray and storing them after draining. Drain the water from the pressure switch tube as well.

Drain down any solar panels, and leave the plumbing fittings or gate valves open to the atmosphere. Even in winter the heat in a solar panel can be intense, and expanding air can be as hazardous to the panels as freezing water.

Be sure the lines between the equipment are empty or at least filled with antifreeze. Remove any spa jet booster pumps and store them. Leave all valves open and disassemble any three-port valves. Finally, if the spa has a fill line, shut off the water supply and drain that line as well.

8. **Extra Precautions** Put a little dirt or gravel in an empty plastic bottle and leave it in the skimmer. The weight will help it stay upright in the skimmer if rain or snow refills it with water. Should the water

subsequently freeze, the ice will compress the bottle, not crack the skimmer. Similarly, leave some plastic milk or chlorine jugs floating on the surface of the spa. Fill them about one-fourth with water so they will "grip" if the water freezes and not just pop out onto the surface of the ice. Again, the goal is to create an expansion joint so that the ice will crush the jugs, not the spa walls.

If there are exposed pipes, which you suspect may still contain water, in the equipment area, under decks, or in self-contained spa cabinets, wrap them with insulation tape (a good idea any-way, to reduce heat loss when operating the system). In extremely cold areas, wrap them with electrical insulation tape, available at most hardware stores on a seasonal basis. These tapes actually warm the pipe with a low-level electric current, plugging into wall outlets.

The air blower ring must be drained also. When the spa is empty, use the blower to evacuate the line. If you need to refill the spa, pour antifreeze over the air holes so it penetrates the openings. Then dry the surface and apply duct tape over the holes before refilling the spa. When you reopen the spa, you will probably need to clean the tape adhesive off with acetone, but the extra effort is better than finding a leaking spa.

If metal rails and light fixtures cannot be removed, protect them against corrosion with a coat of petroleum jelly.

9. **Cover the Spa** Some cover is better than none, as it will inhibit algae growth and keep heavy debris out of the spa. Sheet vinyl covers (Fig. 8-8) are very inexpensive and can be held in place with rocks or sandbags around the edge of the spa. Put your expensive foam or custom cover in storage to protect it over the winter too.

10. **Shut Off Access to the Spa** Whether it is a commercial or residential spa, block-ing access to the closed installation is an important safety precaution. Yellow caution tape strung around the spa, locked gates and fences, and extra signage will keep the spa from becom-ing an inadvertent hazard and will limit your liability.

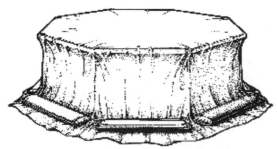

FIGURE 8-8 Vinyl winter cover for spa.

11. **Clean Up** Take this opportunity to properly dispose of any extra chemicals or test kit reagents that won't be used during the winter and will not be potent in the spring. Soda ash and acids are about the only water chemicals that will still be good after prolonged storage, so make sure they are packed in watertight containers and stored in well-ventilated areas away from water or heat sources.

During the winter you will still need to check on the spa. Snow or rain may have raised the water level or sunk the cover. Animals or heavy debris may have fallen in the spa and would be better removed now than in spring.

Reopening the spa is essentially the reverse of the shutdown procedure, with emphasis on balancing the water before restarting the circulation system.

Installation and Special Repairs

Most of the book thus far assumes you already have a spa or hot tub. But what if you are thinking of adding one to your home? What about improving the appearance of an older spa? What if your spa or hot tub develops a leak? This chapter will provide a practical guide to the choice of do it yourself or hire a pro for each of these questions.

Installing a Spa

In this section I will deal with the installation of a shell-type spa—fiberglass or acrylic. Jetted bathtubs are installed like any other home bathtub, with the addition of the jet pump, plumbing, and wiring.

This section assumes that you have chosen a style of spa and a location. If that is not the case, some guidelines that might help in this process include the following.

The Shell

How many people will use the spa at any time? Spas are sold as two-person, four-person, etc., indicating the number of bodies the shell can comfortably accommodate (Fig. 9-1A and B). The molded-in seating or lounges (Fig. 9-1C) will make that determination fairly obvious. It is important to consider your priorities as well. For example, you might have one or two parties each summer when 8 or 10 people will use the

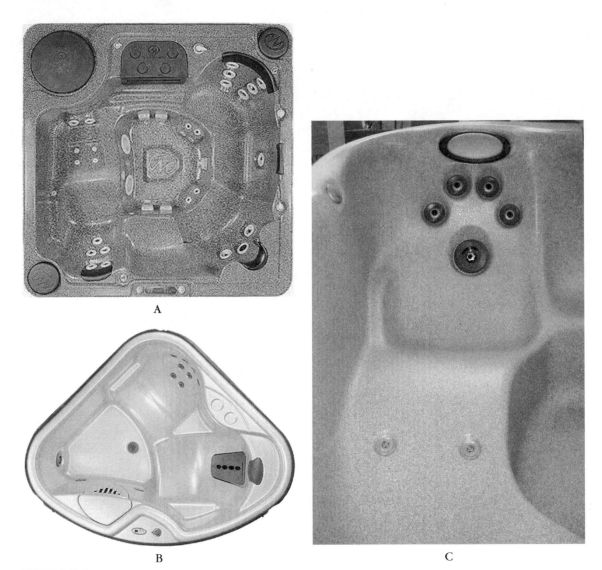

A

B

C

FIGURE 9-1   (A, B) Spa shell. (C) Contoured spa seating. *A, B: Master Spas.*

spa, but the remainder of the year it will be used by only two people. The amount of space the 10-person spa requires on the deck or in the yard (Fig. 9-2) and the annual costs of maintenance, heating, and electricity might not make sense if it is used by only two people.

Consider the manner in which the spa will be used. You will want straight bench seating if it is used mostly for socializing with several people, but will prefer lounges if it is used for therapeutic bathing.

FIGURE 9-2 Backyard deck location for spa installation. *Bradford Spas.*

Note also the size of the most frequent users. I was once asked to install a spa for a very famous actor/musician in southern California who is very short. His builder had already purchased the shell, pre-plumbed with skimmer, drains, and jets. When the project was complete, this person sat in his new spa and found the water level up to his eyeballs. All the spa seating had been designed for a taller person whose upper body length would leave the shoulders at the waterline. Most spas have seats of varying depth to accommodate a variety of human proportions, but this one hadn't counted on its only occupant being so short. The solution was to lower the skimmer so the waterline would be lower, something we should have thought about first.

The Plumbing

Similar considerations need to be taken into account when plumbing the shell. Depending on the intended users, you might be able to install a shell that has been preplumbed to accommodate the average bather (Fig. 9-3). Typically two jets per person are included in a spa, set in the walls at varying heights. An air bubble ring is also provided to create general turbulence in the water for a massaging effect.

If the users have specific preferences about number, type, and location of jets, it makes more sense to buy the shell and plumb it to their

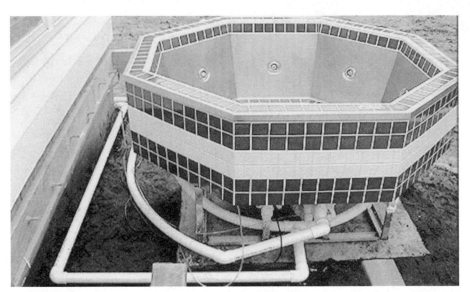

FIGURE 9-3 Preplumbed spa shell. *Bradford Spas.*

specifications. By inquiring at your supply house, you can make a list of options available, such as small pinpoint jets, larger standard massage jets, rotating jets, and those that can be turned on and off to put more or less pressure in one specific area of the spa. Some people only care about the general turbulence of the hot water and require one or more air bubble rings, while others are only interested in the massage jets and don't want to invest the extra money in the air bubble ring and electric blower required to power it. A blower can also be plumbed to turbocharge the jets. In short, familiarize yourself with your actual preferences and design the spa accordingly.

Finally, price makes a difference to almost every consumer. The self-contained spa, built with the same shell and plumbing you might provide for a custom design, will cost anywhere from $2000 to $5000. By the time you factor in the installation of the shell, plumbing, equipment, and subcontractors, you can easily spend up to 10 times that for a built-in spa. Before investing a great deal of time talking about custom options, consider your budget for the job.

The Equipment

At the same time as you are choosing the number and styles of plumbing options, you need to review the equipment requirements of each choice.

The pump needs to be of sufficient strength to provide adequate power to the jets. The general rule of thumb is that each jet needs ¼ horsepower, so a 2-horsepower pump could theoretically give adequate performance to eight jets. Remember also that the turnover rate must be satisfied and that most jets, to operate as designed, require about 15 gallons per minute (57 liters per minute) of flow each.

Using those general guidelines, decide on the size of the pump needed and whether more than one pump is required. In larger installations, you might want to divide the spa into two halves and operate each side from a separate pump. When making pump selections, consult your electrical subcontractor to determine that adequate electricity is available for your choices and understand the costs of any upgrading that might be required of the electrical breaker panel.

Cartridge filters are the preferred choice for spas, following the selection and sizing criteria outlined previously. Consider the volume of use when choosing a filter for a spa. If the customers will use it frequently (many people jump in the spa when they get home from work before greeting the rest of the family), choose a filter size larger than actually needed to reduce the frequency of required teardown and cleaning. If you are using a regular cleaning service, you might be less concerned about the frequency of teardown than if you plan to perform your own routine maintenance.

The most important consideration in choosing a heater is the intended use. If the use is unpredictable, you will need a larger heater to raise the temperature quickly. If you know the spa needs to be heated daily at 6 p.m., you can choose a smaller heater and set the system to turn on with a time clock early enough to reach the desired temperature by the desired time.

Consult with your plumbing contractor about the availability of gas supply, and consider the cost of upgrading lines and meters and adding the supply line to the equipment location. If there is no gas available in your area, consult with your electrician about adequate electrical supply for an electric heater. Since even the largest electric heaters take a long time to heat a spa, choose the largest one your electrical supply can afford.

Decide if you want one or more blower rings, which force bubbles into the water for a general massage. Rings are often located in the floor and/or seats of the spa. You might add a second blower to turbocharge the jets, or install a three-port valve to divert the air flow of a single blower to the air ring and jets. Again, consult with an electrician about available electrical supply and the cost of wiring to the equipment location.

To provide a true overall estimate of the job, make a list of additional equipment you might want to consider.

- Cover
- Chlorinator
- Lighting
- Mist spray
- Fill line (automatic or manual)
- Control system (electronic, manual, or air buttons)
- Time clock (might be included in the control system)
- Handrails, built-in drink holders, and padded neck rests

The supply house in your area or the Web can provide catalogs of available products to help you make these choices.

Spa Installation: Step by Step

RATING: PRO

Once you made the decisions about what to install, the following list outlines the steps required.

1. **Planning** Obtain all necessary local building permits prior to starting the job. As in other jobs requiring construction methods beyond the normal work of the water technician, I prefer to subcontract the work to others and let them do what they are trained to do, while I install the shell, plumbing, and equipment. In addition to hiring a general contractor to work on the construction and permit portions of the job, you might want to subcontract moving the electrical, gas, or water supply connections to the equipment area. You might also hire landscaping and masonry subcontractors.

2. **Foundation** Working with your general contractor, carpenter, mason, or other subcontractors, plan the site preparation and coordinate the spa installation with any deck or other foundation work.

Whether you will be preparing the deck and foundation or the subcontractor is doing it, a few concerns are often overlooked. First, be sure the deck is level. This might seem obvious, but if there is the slightest tilt, it will become very apparent when the spa is filled with water. A spa that is sitting on a deck that is not level will have a waterline that is high on one side and low on the other. This could be important if, for example, the high end is the side of the spa with the skimmer. To keep enough water in the skimmer to keep it from running dry, it might be necessary to fill the spa to the point of overflowing at the low end. Remember also that people in the spa will displace a great deal of water, so even if the variations are minor, they will become significant when the spa is in use.

> **TOOLS OF THE TRADE: SPA AND HOT TUB INSTALLATIONS**
>
> - Shovels
> - Level
> - Concrete, sand, mixing bin
> - Plumbing tools (see Chap. 2)
> - Felt-tip marker
> - Electric drill, bits, extension cord
> - Electric jigsaw
> - Screwdriver
> - Channel lock–type pliers
> - Silicone sealant
> - 2$\frac{1}{4}$-inch (57-millimeter) hole saw
> - Needle-nose pliers
> - Wire cutter/stripper

Of course you might start out with level surfaces, but if the foundation for the spa and deck is not stable, the spa might not remain level. A spa that is full of water and three or four people will weigh over 2 tons (1800 kilograms). In a relatively small area, that weight creates a great deal of force for movement if it is not absolutely secure.

You will need to follow the building code in your area, but the objective is to secure the spa from shifting by supporting it in as many locations as possible, evenly set on the deck. Figure 9-4 shows how the spa lip rests on the deck. It also shows how the spa rests in the excavation, surrounded by sand for general lateral support. The level is maintained by the deck.

Obviously, if the spa shifts the sand will compress and fail to support the spa. The advantage of that is that the sand also gives enough to prevent damage to the plumbing.

Once the location of the spa is determined, a simple hole is dug in the ground, large enough to accommodate both the spa and the

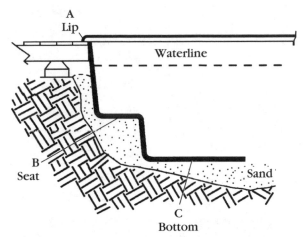

FIGURE 9-4 Spa foundation with sand backfill.

plumbing. Figure 9-4 shows how the excavation is filled with sand, but remember, you might need to excavate the sand at some later time to repair leaks, so dig a hole that is also large enough to accommodate a technician making repairs or inspection.

For this reason, I prefer an installation that supports the spa without filling the excavation with anything. As long as the base of the spa is supported (Fig. 9-5) and the lip is secure on the deck, there is no need to cover the plumbing with sand. Believe me, at some point during the life of the spa you will need to access the plumbing, so preparation for that now will pay dividends later. Your subcontractors can help you install simple post and pier supports (Fig. 9-5A) to create a framework to stabilize the spa, leaving the plumbing exposed. If the excavation is large enough and a deck access (or opening in a slope) is created, you will be able to easily diagnose and repair leaks. Be sure to use a solid concrete base in this type of installation (Fig. 9-5B).

You can also prevent leaks in this way. If the spa shifts for any reason, the pressure of the plumbing against the sand can crack joints and fittings. Erosion, earthquakes, and even frequent passing of heavy trucks can cause earth movement and shifting of the spa. My experience is that no matter how well installed it is, every spa leaks sooner or later. Site preparation makes the difference between a simple repair and tearing out the entire installation and starting over.

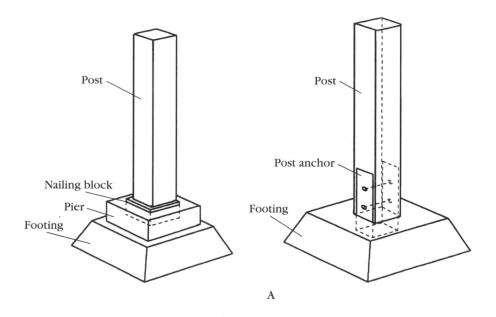

FIGURE 9-5 (A) Post and pier supports. (B) Concrete pad foundation. *B: Bradford Spas.*

3. **Bringing the Spa to the Site** Acrylic and fiberglass gel coats scratch easily, so take care to cover the top of the spa to prevent tools, dirt, or sand from damaging the surface. You will need to work inside the spa, and painters' dropcloths make good working covers.

4. **Plumbing** If the spa is preplumbed, skip ahead to step 7. If not, plumb the spa at the job site. Figure 9-6A shows the plumbing of a typical spa shell. Decide on the number and location of the jets, marking them with a pencil or felt-tip pen. Plan on installing a skimmer even if the spa is indoors in a clean area, because the skimmer will remove body oil or suntan lotion from the surface of the water. Locate the skimmer on the side of the spa closest to the equipment, but consider the seating pattern of the spa and avoid placing it where it will prevent someone from using a lounge or molded seat. Purchase the jet fittings, drains, and skimmer before you cut any holes in the shell.

5. **Installing the Skimmer and Main Drain** As shown in Fig. 9-6B, suction plumbing runs from a skimmer and main drain, joined with a T fitting as shown or an optional three-port valve. Position the skimmer as discussed previously, then determine the appropriate height. The top of the skimmer opening will normally be only 2 or 3 inches (50 or 75 millimeters) below the lip of the spa so that the normal water level, in the middle of the skimmer opening, will be about 6 to 8 inches (15 to 20 centimeters) below the lip. Remember that bodies displace water, so if the skimmer is located too high and the water level is raised to fill the skimmer, when you add bathers the spa will overflow. In addition to the obvious problem of a wet deck and an unsatisfactory installation, the water loss means that when the bathers leave, the water level will drop below the skimmer opening and the pump will run dry. If the skimmer is located too low, the water level will be higher than the skimmer and the spa will have no skimming action. If you are in doubt about the positioning of the skimmer, take a look at a preplumbed spa or a unit already in service to get a feel for the best placement.

The actual installation of the skimmer is simple. The skimmer is supplied with a faceplate and gasket that fit over the opening on the inside of the spa. Using the faceplate as a template, determine the exact position of the skimmer opening by holding the faceplate against the inside of the spa and tracing the opening with a pencil.

A

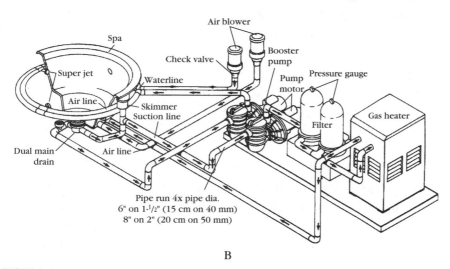

B

FIGURE 9-6 **(A) Plumbing the spa. (B) Typical spa plumbing.** *A: Bradford Spas. B: Sta-Rite Industries.*

Mark the location of the screw holes. Remove the faceplate and drill out the screw hole locations to the size of the screws provided. I drill the screw holes before cutting out the skimmer opening because the spa wall is stronger before removing the material from the opening, and it is less likely to chip or crack than if I were to drill the holes after that portion was removed. Next, drill a starter hole within the perimeter of the skimmer opening to create an access for your saw blade. Use an electric reciprocating saw, jig saw, or manual keyhole saw (if you have a lot of patience) to cut out the skimmer opening. Hold the skimmer against the back of the spa, and apply the gasket and faceplate to the interior side of the spa. Screw them together with the hardware provided. Apply some silicone sealant to the bolts before slipping them through the holes. When you tighten the bolts, the sealant will fill any gaps.

Install the main drain, following the procedures described in step 6 for jet fittings. Some building codes require two main drains attached to the same suction line, joined by a T fitting or one drain with two suction ports, as shown in Fig. 9-6B. Since spa suction is extremely strong, should one drain be covered by a hand or foot, the other one will then pull water rather than a single one that might injure the bather. In this case, the second drain is usually mounted in the side while the primary drain is located in the floor of the spa.

Install any light fittings, following the instructions with each different model. Some will require creating a light niche in which the actual fixture is mounted. The niche is installed on the shell in the same manner as the skimmer. Others are designed more like a main drain, where the fixture is threaded to fit through a hole in the spa wall and screws into a body on the exterior like a jet fitting.

6. **Plumbing the Jets** A typical 1½-inch (40-millimeter) plumbed jet fitting requires a hole in the shell of 2¼ inches (57 millimeters) to accommodate the throat of the jet (Fig. 9-7A). Using your electric drill and a 2¼-inch hole saw, cut holes for the jets in the locations marked. The jet body slides through the hole, and a nut on the back secures it in place. Be sure to use the gasket provided on the inside (the water side) of the fitting to prevent leaks. Tighten the nut with channel-lock pliers, applying enough force to tighten but not crack the plastic nut.

The other type of jet uses a jet body that is held against the exterior (back side) of the spa, while a threaded throat passes through the hole in the spa wall and screws into the body. Jet wrenches are provided by the makers of these jets for tightening. On the face of the throat you will find two holes. The wrench fits into these holes to grip the throat fitting. While holding the jet body (usually requiring a helper), tighten the

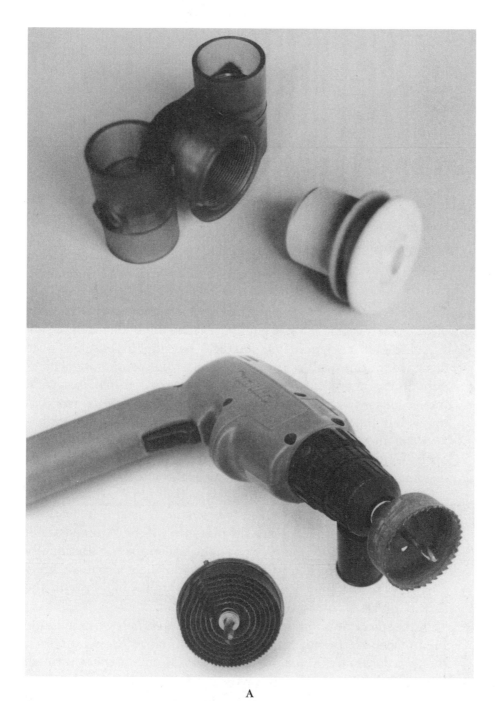

A

FIGURE 9-7 (A) Jet fitting and hole cutting bits. (B) Spa jets installed. (C) Finished spa installation. *C: Bradford Spas.*

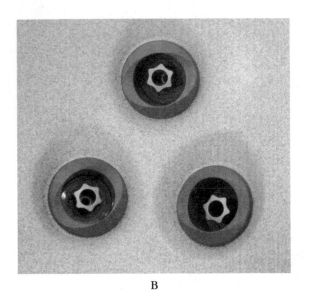

B

FIGURE 9-7 *(Continued)*

throat fitting into the body. Use enough force to tighten the throat against the gasket for a watertight seal, but not so much that you crack the fitting. Make sure to install all jet fittings with the water plumbing below and the air plumbing on top.

Some manufacturers make micro-jets (Fig. 9-7B), smaller versions of the 1½-inch (40-millimeter) standard jet, built to accept 1-inch (25-millimeter) PVC plumbing. These are most often used in jetted bathtubs, but some people prefer the microjet in combination with the standard jet because you can use more of the microjets with the same equipment than the standard units.

Whichever design of jet you use, be sure to buy the ones designed for thin-wall spas. Jets are also made with identical components for hot tubs, except that the threaded throat is several inches longer to accommodate the greater thickness of the wood as opposed to the thinner fiberglass or acrylic walls (Fig. 9-7A, top).

Using flex PVC and the plumbing techniques described previously, connect the jets as illustrated in Fig. 9-6B. If you are using one pump, the return line that sends water through the jets should be diverted with a T fitting, sending one-half of the flow to one-half of the jets and the other half to the remaining jets. If instead the flow runs the length of the jets, the pressure will be extremely high for the first few and extremely low on the last one or two. By equally dividing the flow you avoid that problem. You might also use a three-port valve instead of a T fitting, so that you can equally divide the flow or divert all of it to one side or the other.

If the spa is large and more than one pump is being used, divide the number of jets into two groups and plumb the flow from each pump accordingly. Since one pump will circulate water through the filter and heater, plumb it with fewer jets. For example, if the spa has 10 jets, plumb the filter and heater pump with four jets and the unrestricted second pump with six jets.

C

FIGURE 9-7 *(Continued)*

In each case, cap the dead (terminal) ends of each jet line with 1½-inch (40-millimeter) PVC caps attached to a short length of flex PVC pipe. I prefer this method to inserting a 1½-inch PVC plug into the jet, because if the plumbing needs to be changed for any reason, the plug makes that side of the jet permanently inaccessible. If you do need to use the other end of the jet that is capped, you can cut off the cap and glue on a slip coupling, plumbing anything on from there.

Connect the air side of the jets together. If the jets are to be turbo-charged from a blower, plumb them as the water side is, with a T fitting or three-port valve between equal numbers of jets. Cap the dead ends as described previously.

If the jets will be aerated instead from the atmosphere, there are two choices. First, on the top of each jet is a ¾-inch (19-millimeter) opening to accommodate a length of pipe that extends up to the lip of the spa, aerating each jet individually. The other choice is to glue these openings shut with the plugs provided and stub one or two common air lines to the lip of the spa, as shown in Fig. 9-6B. With this method, an air control fitting can be mounted in the lip of the spa to which the common air pipe is glued. This air control fitting allows the spa user to close off all air supply, resulting in a jet that moves only water, or the air supply can be opened for the desired amount of added aeration. If the spa has been constructed with one or more air blower rings, they will be provided with a 2-inch (50-millimeter) stub in which to glue the PVC pipe.

7. **Testing** Before installing the spa, perform a leak test on the plumbing. If the spa is preplumbed, it is important to perform this test to be sure the factory did a good job and that nothing was damaged in transit. It is much easier to repair or replace the plumbing above ground than to wait until the unit is installed. Glue flex PVC into the T fittings or three-port valves, and run enough length to bend the pipe over the top of the spa. You will have one line for the main drain, one for the skimmer, one for the air ring, and one for the return water to the jets (or two if you have installed a two-pump system). Cut the PVC and tie the lengths above the top of the spa, then fill the spa with water. There should be no leaks before you set the spa in the ground. Let the water sit for an hour or more to be sure there are no slow leaks. Take care when filling not to wet the outside of the spa or the ground, making it difficult to observe any leaks. If you have the equipment, pressure-testing the spa is a good idea (and might be required by building codes in your area), although I have found that if the plumbing is done with care and the spa passes this simple fill-up test, it will not leak under pressure. When you are sure there are no leaks, drain the spa.

8. **Setting Up** Set the spa in the ground, onto the deck and any other supports. Carefully lower the plumbed spa into the ground, using enough helpers to avoid dropping the unit or resting it on the fragile plumbing. If the spa is to be backfilled with sand, once the unit has been stabilized and secured, perform another leak test. This might seem redundant, but believe me it is easier to spend an extra hour at this point than to make repairs later that require excavating the sand to find the leaks. If you took my earlier advice and used supports, you won't need to leak-test at this point, because without the sand, you will be able to see if leaks are present once the job is complete.

9. **Setting the Equipment** The equipment should be located as close to the spa as possible to avoid pressure loss in the system and to avoid heat loss by running water through long pipe runs. The level of the equipment should be as near the water level of the spa as possible to avoid priming problems. Working with your sub-contractors, prepare the site with a concrete pad of adequate size to accommodate the equipment, leaving enough room between components to allow access for repairs or maintenance. If you need to provide your own equipment pad, follow the instructions in the section on foundations for hot tub installations. Complete

---

TRICKS OF THE TRADE: SPA INSTALLATIONS

Many technicians and do-it-yourselfers are intimidated by the idea of installing a spa, but it's actually easy. In essence, spa installation is a combination of activities described throughout the book. Here are some tips to make the job even easier and more successful:

- In preparing the site, look around for drainage patterns. Try to contour the site to keep heavy rains from flooding (and perhaps undermining) your work. If water collection in the pit is unavoidable, install an automatic sump pump.

- Look around for potential future problems from tree roots.

- In cold climates, plan to both insulate and bury pipes between equipment and the spa.

- Consider the prevailing winds (especially at the coldest season in which the spa will be used each year). If possible, plan landscaping or decorative fencing to shelter the spa from cold winds that will chill both bathers and water.

- Consider safety—can the spa be accessed by children and pets? Plan to add fencing or locking safety covers.

- Plan the steps between the house and spa. Many enjoyable trips to the spa are rendered unpleasant when bare feet traverse gravel or splinter-covered areas on the way back to the house. Be sure that stepping stones or paths installed are textured to avoid slips and falls.

- At some point, you will need to access the plumbing and equipment again for repair or replacements. Leave adequate room around the spa for this purpose. If you are filling in the space around the spa and below a deck, use sand or other materials that are easily excavated later. Don't ever backfill these voids until the spa and equipment have been repeatedly and thoroughly tested for leaks.

the gas and electrical supply connections to the equipment, and plumb the components using the methods described in previous chapters.

10. **Plumbing** Trench as needed to run plumbing and electrical lines between the equipment area and the spa. Plumb the equipment and spa stub-outs together. If there are long runs of above-ground pipe under a deck or in other exposed areas, wrap them with insulation to prevent heat loss.

    Special attention should be given to the blower lines to ensure that water doesn't travel back up the line into the electric motor. See the section concerning blowers for details. Before closing the trenches, connect any control devices or switches planned for the installation, such as air switch buttons with air hoses running to the equipment area through a pipe in the trench.

11. **Electrical Requirements** Table 9-1 describes requirements for portable spas, but they are good general rules of thumb for inground installations also (except that the equipment for inground units will probably always be hard-wired instead of relying on a plug-in equipment package and electrical receptacle).

As with any installation procedure, you will encounter variations in the spa installation process. Nothing in this procedure is set in stone,

TABLE 9-1 Spa Electrical Supply Guide

	120-volt installations	**240-volt installations**
Wire	3-wire 20-amp; GFI	4-wire 50-amp; GFI
Plug-in cords	15-foot (4.5-meter) max. (hard-wired over 15 feet)	Not allowed
Other receptacles nearby	Within 10 feet (3 meters) must have GFI	Within 10 feet (3 meters) must have GFI
Bonding	All metal within 5 feet (1.5 meters) bonded to equipment ground grid	All metal within 5 feet (1.5 meters) bonded to equipment ground grid
Circuit breaker	20 amps	50 or 60 amps

and each step requires planning and creativity to complete the job in a way that allows for correction of problems or later repairs. Remember, it is much easier and less expensive to change the design on paper than in the field. Planning and forethought will make a big difference in your satisfaction with the final results (Fig. 9-7C).

Draining a Spa or Hot Tub

RATING: EASY

Many of the procedures outlined in this chapter require draining the spa. It may seem obvious that you open a drain valve in the equipment area or use a submersible pump to do the job, but this simple task can create many problems if you don't take all factors into consideration. Therefore, before we present information requiring this procedure, here are the concerns about draining.

1. **Shutdown** Turn off all circulation equipment at the circuit breakers, so there is no chance it will start up from a time clock. Be sure that underwater lights are also switched off and not connected to a time clock. Often there are switches inside the home, so tape over these with a note to make sure all family members know there is a reason to keep that switch off.

2. **Safety** Especially at a commercial spa, run yellow "Caution" tape around the deck to keep unwary visitors from falling in the empty spa. Turn deck furniture on its side and use it as a physical barrier as well. Post signs at every location to the deck about what is going on and that the spa will be closed for several days. It may also seem obvious, but with today's busy families, everyone may not have been told that the spa was going to be drained, so even on residential jobs, put up signs, tape, or barriers.

3. **Drain the Spa** The easiest way to drain a spa is to open a drain valve in the equipment area and let gravity do its job. If the equipment is above the lowest level of water in the spa, this method will not fully drain it. In that case, use a submersible pump.

    When draining a spa with your submersible pump (Fig. 9-8) and hoses, direct the flow to a deck drain if possible. This will send the wastewater through an intended channel, rather than over a backyard garden or down a hill where erosion damage can occur from

FIGURE 9-8   Submersible pump for draining a spa.

such a large volume of fast-moving water. When you have started pumping, however, watch the flow into the drain for several minutes. Sometimes debris will back up in the line and it will overflow, but not until it has filled several hundred yards of pipe. It may take time, but the clogged drain will flood backyards and living rooms and may actually flush the water back into the spa when you have gone.

If deck drains cannot accommodate the flow, connect several vacuum or backwash hoses together and run the wastewater into the street where it is carried to storm drains. Again, watch the flow and make sure after a few minutes it has not backed up a storm drain and begun to flood the street.

In areas where water conservation is a concern, drain the spa when the chlorine residual has gone below 1 ppm, then let the flow irrigate lawns and gardens. Another technique along these lines is to punch holes along the length of a backwash hose (or old vacuum hose) and seal up the discharge end (tie a knot in a backwash hose, or use plumbing fittings to close the vacuum hose). This acts as a huge sprinkler, evenly distributing the water across lawns and gardens.

One other word of warning about deck drains. They are usually made of PVC, but since they don't carry water under pressure, they are not usually pressure-tested. If ground movement or other erosion has destabilized them, the pipe may have separated and much of your wastewater will end up eroding the backfill around the spa. If you have any doubts about the integrity of the deck drain or its ability to handle the water, run hoses into the street.

In some jurisdictions there may be restrictions on pumping out pools and spas relative to the permissible volume and even the permissible chemical makeup. Extremely low pH water may have to be neutralized before you pump it into municipal stormwater or sewer lines. Check your local codes before turning on the pump.

TRICKS OF THE TRADE: SUBMERSIBLE PUMP SAFETY

1. When lowering a submersible pump into a spa, never do so by the cord. Attach a nylon rope to the bracket or handle, and lower it that way to avoid pulling electrical wires loose. Not only will the pump fail if the wires come loose, but also when you plug in the cord, it may electrify the water.

2. Buy a ground fault circuit interrupter (GFI) which can be plugged into the wall socket before plugging in the cord of the pump. It will be the best $30 to $50 you ever spent. I have been in water with a pump that had a slight short and felt the "tingling" of electricity conducted through water, and I have known technicians who have been killed by not taking this aspect of operational safety seriously.

3. Rig a remote on/off switch that plugs into the socket first as well, allowing you to operate the pump when you are working in the spa.

Repairs and Remodeling

Modern materials used to make today's spas are virtually indestructible, so an old spa can be repaired or remodeled to make it appear and function as if new.

Spa Leak Repair

To fix a suspected leak, first you have to locate it. Leaks are usually signs that something else has gone wrong. For example, a cracked pipe may point to a shifting foundation of the spa itself. Moreover, leaks often come in bunches. When you suspect a leak, look for all possible locations and determine why the leaks occurred in the first place (Fig. 9-9).

The first place to look for leaks is in the exposed equipment and plumbing. If that fails, there are three basic ways to find hidden leaks that are available to do-it-yourselfers and two more ways that typically require professional help.

EVAPORATION TEST

RATING: EASY

The simplest method of leak detection is to fill a bucket and place it on the deck adjacent to the spa. Mark the level in the bucket with an indelible felt-tip marker, and do likewise for the water level in the spa. Turn off the spa circulation to eliminate any variables in evaporation.

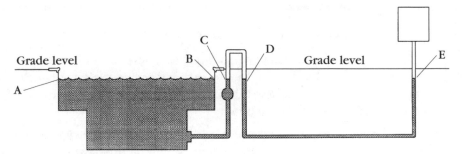

FIGURE 9-9 Finding a leak in a spa.

After several days, mark the new level of water in the bucket and spa. They should have evaporated an equal number of inches (or millimeters). If the spa level has lowered significantly more, there is likely a leak. If both vessels have lowered a similar number of inches, then there is no leak.

DYE TESTING

RATING: ADVANCED

Perhaps the simplest method of detecting a leak of the vessel itself is to "shoot" a dye test. As the name implies, a colored dye is disbursed in suspected areas and as the dye disappears, the leak is found.

1. **Prepare** Clean and brush the spa thoroughly. Cracks can sometimes be hidden by dirt or other material settled in the crack itself. Pay careful attention to steps, corners, and around fittings.

2. **Inspect** Turn off the circulation and begin the examination on a calm day. Wind rippling the surface will make it difficult to see small cracks. You may also want to squirt a little tile soap across the surface to sharpen visibility further (and you may need to repeat that process from time to time during the exam). Examine the spa for obvious cracks, beginning with the tile line. You may not need to go much further if you see gaps or missing tiles. Tap the tiles with the handle of a screwdriver (or gently with a hammer) to see if any fall off, are loose, or sound hollow. Note any "positive" locations. Continue to visually inspect the interior surfaces, looking for cracks or discolored patches of plaster (or plastic in the case of acrylic spas), noting anything you suspect.

3. **Apply the Dye** The dye test itself can easily be conducted by using an old test kit reagent bottle or similar squeeze bottle filled with food dye (available at any grocery store). Some technicians use phenol red (from their water testing kit), but you will need to check many locations and it is unwise to inject that much acidic chemical into the water. Work around the spa, particularly the locations of suspected leaks from your exam. You will need to get into the spa, including reaching down to the main drain, to do a thorough job. At each crack or suspected area, aim the nozzle of the bottle at the crack. Squeeze a bit of dye into the area and watch it. If the dye simply swirls around the crack without being sucked in, then there is no leak in that area. If, however, the dye is sucked into the crack, it is riding on a flow of water leaking from the spa. The speed with which the dye disappears will help you estimate the size of the leak.

4. **Find All the Leaks** As with the visual inspection, continue around the entire spa, looking for leaks, to exhaust all possibilities. Be especially careful around skimmers, steps, rails, ladders, or other fittings. Light niches are often the source of unexplained leaks, so pay close attention to the entire light fixture area. If you are having an especially difficult time finding a leak, you may even wish to remove the light fixture and dye-test directly into the niche itself, concentrating on the area where the cable passes through. Don't forget the interior of the skimmer and the main drain as well as return outlets.

When you have thoroughly examined the spa with the dye test, you will know what repair problems you are faced with.

DRAIN-DOWN TEST

RATING: ADVANCED

If you have tried the evaporation test and dye testing and still not found the leak, try the drain-down method.

1. **Prepare** Turn off the equipment and mark the level of the water in the spa.

2. **Mark** Mark the level again at the same time each day to establish a rate of leak. Because of normal evaporation, the level will continue to decrease indefinitely; however, the objective is to determine when the level stops lowering as a result of the leak. If, for example,

you record a loss of 2 inches (50 millimeters) per day for 4 days, then the rate slows to 1 inch (25 millimeters) every 5 days, you will know the level at which the rate slowed was the level of the leak. Mark that level of rate transition.

3. **Inspect** Examine all possible leak areas along the transition level. The leak must be along this line. For example, if the water loss slowed when it reached the level of a particular return outlet, you might reasonably suspect the leak to be in that plumbing line. If the water slows when it lowers to the area of the light niche, the leak will likely be in there.

The only fault of this method is that it is an indicator, not a precise tool. Since water seeks the same level in all plumbing and parts of the vessel, the water may stop at the level of a certain plumbing fixture, but the leak may actually be in an entirely different location which is coincidentally at the same level (Fig. 9-9). In any case, the new level will tell you where to look further and where you need not look.

LEAK DETECTORS AND PRESSURE TESTING

RATING: PRO

When the above methods fail to help you locate the leak or you wish to further verify your assumptions, there are two other detailed methods of leak location which usually require the services of a pro.

There are electronic listening devices called *geophones* that can actually "hear" water dripping or flowing. By applying such devices around the spa and related plumbing, an operator can identify where water is moving out of the system. Because these devices are expensive and their operation requires a great deal of experience and skill, most service technicians don't buy or use them. There are numerous professionals who do, however, and are easily found in the phonebook or through referrals at your supply house.

The second method used by other professionals to help you find your leaks is pressure-testing equipment. It is not difficult to pressure-test a plumbing system with the knowledge already presented in this book. However, the amount of time and additional equipment (plugs, adapter fittings, compressed air, and related fittings) makes this type of testing impractical for most homeowners. The companies that conduct leak testing may also conduct pressure testing.

Repairing Leaks

Now that you have determined where the leak is located, what's next? Repairs can be fairly simple or require extensive disassembly of the spa and its surroundings.

If the jets are found to be leaking around the faceplate, they probably need to be tightened. Use your jet wrench to check each fitting. Time, harsh chemicals, and heat can deteriorate the gasket between the jet body and throat, so tightening might not seal the leak. Unfortunately, you can't usually replace gaskets. If you remove the throat, the jet body behind the spa will invariably shift slightly, making realignment for reassembly impossible. If you can access the jet body, replacement of the gasket is possible.

If the gasket seal is found to be leaking, try draining the water level down below the jets and unscrew the throat slightly. Allow the area to dry completely. If the gasket is brittle and/or extremely deteriorated, remove it. Take a new gasket and cut it along one side so you can open it and slip it around the jet throat. Obviously, the place where you cut the gasket will not seal when you tighten the throat down, so fill the gap with silicone sealant, then tighten. Wipe the excess sealant that oozes out as you tighten away from the face of the jet. Wipe in an even, steady motion, forcing the material into the seal area and smoothing it over the joint of the jet face and spa wall. The objective is to form a watertight seal and thoroughly cover the work with the sealant. Allow the work to dry completely before refilling the spa and dye-testing again.

Similarly, the integrity of other joints, such as the skimmer, drains, and light fixtures, is often compromised. In each case, tighten whatever you can and then seal the entire assembly with silicone sealant. These seals will often leak again as the silicone shrinks from constant temperature extremes and harsh chemicals, but this provides an inexpensive and quick way to extend the life of the spa. Of course, the best repair is to completely disassemble the leaking component and replace its gasket or the component itself. But if the spa installation does not allow that as an option, silicone is your best bet.

If you have made these repairs in numerous places in a spa, you need a complete overhaul. In this case, you remove enough of the deck and backfill to expose the plumbing, replace the fittings and fixtures in the spa, then rebuild the assembly. These jobs can be very labor-intensive. I usually advise the customer that the charges will be actual time and materials, at a preapproved labor rate and parts estimate, with a probable total but no guarantee. The advantage of the rebuild is that the customer doesn't keep throwing $100 repairs into the spa for prolonged periods. One other choice is to fiberglass the leaking fitting(s) to the shell as described later.

CRACKS

The other type of leak that a spa is subject to is the crack. Earth settling or shifting, kids jumping in the spa, and age can create cracks in the spa. Acrylic spas can be repaired as described below, and fiberglass spas can be repaired with the techniques described in the hot tub section.

PATCHING AND REPAIRING SPA SURFACES

If your spa is gunite and plaster, you might need to be familiar with simple patching techniques applicable to the walls. This section will also examine repairs to acrylic surfaces.

Plaster (RATING: ADVANCED):   When the spa is drained for any reason, you have an opportunity to look for plaster blisters or pop-offs, areas where the plaster has come away (delaminated) from the underlying surface. Such areas can be anything from the size of a quarter to dinner plates to several square feet. Beyond that and you might need to replaster.

The causes of delamination, also called *calcium bleed*, are numerous. The original plaster job might have had poorly prepared surfaces, poorly mixed materials, or too much drying before the water was added. Most likely, however, improper water chemistry has created aggressive water, stripping calcium from the plaster, weakening the plaster layer and allowing it to separate from the underlying surface.

Some blemishes will be obvious, the plaster having cracked and fallen away. Others will appear as discolored spots from the water that has entered between the plaster and the subsurface. You can also look for plaster faults by tapping suspected areas with the handle of a screwdriver or with a hammer to listen for hollow sounds (assuming the spa is empty). If you are committed to making repairs, poke around the entire spa with a chisel or screwdriver to pop the blisters. They have to come off anyway, so you might as well discover them all. Once you've been around the entire spa, you can then evaluate if patching is the answer or if replastering is more logical.

When you have identified blisters, chip the loose plaster away all around the area until you reach solid, dry plaster. Use a hammer, chisel, or screwdriver depending on the size of the blister. Once it is completely exposed, clean the area of all water and loose debris. To make sure the final patch blends in and appears even, clean up the jagged edges of the blister area by sanding the perimeter. The most effective way to smooth out these edges, especially if there are many blisters to repair, is to use a power grinder with a small [6-inch-diameter (15-centimeter)] diamond grinding wheel. The objective is to create a clean, smooth edge all around. You are now ready to patch.

The best way to learn how to work with any repair material is to practice in the workshop before attempting the repair in the field. A little wasted material is less costly (in time and money) than making mistakes on your spa. All these materials are available at your supply house.

Mixing a batch of plaster for patching (or practice) and the actual application are simple.

1. **Water** Start with a clean 5-gallon (19-liter) bucket, adding 4 cups (1 liter) of water.

2. **Mixture** Slowly stir in calcium chloride (the powder used to make the mixture set up) until the liquid becomes warm to the touch. The reaction of the calcium chloride will actually make the water heat up. You will use less than 1 cup (250 milliliters), but keep monitoring the temperature (the chemical will not harm your fingers). Too much calcium chloride can actually melt plastic buckets, so don't be too heavy-handed. The general rule of thumb is 1 part calcium chloride for every 10 parts of cement used. You can also make patch material without it for a slower-drying mixture. This might be helpful if the area to be patched is not underwater and it is a hot, dry day (or a windy day when the material will dry quickly anyway). Faster drying means greater likelihood of cracking. Again, practice with several mixtures and compare results so you become familiar with both the art and the science of plaster patching.

3. **Admix** Add 8 cups (2 liters) of white cement and 4 cups (1 liter) of sand (the supply house will have a specially mixed grain size of sand called *aggregate*). Mix the materials thoroughly with your hands, adding water if needed to achieve the texture desired for the particular patch job.

TRICKS OF THE TRADE: PATCH MIXING

- An alternate method of mixing the patch material, which is especially useful if you are preparing a very dry mix for an underwater patch, is to first combine the dry materials in a bucket, then slowly add the water until the desired consistency is achieved. If you dry-mix, first put the sand in the bucket, then stir in the calcium chloride and cement (because these are lighter than the sand, they will combine more readily if they are added to the sand, rather than the other way round).

- For patching vertical surfaces, pool walls, for example, the mixture should be thicker than if patching a horizontal surface. The texture of bread dough will make the patch material adhere and stay in place better as it dries. The looser texture for horizontal surfaces gives greater flexibility and drying time when making the repair.

- When patching underwater (which is most common), the texture should be as dry as you can make it and still work the material. Obviously, as you fill the patch area underwater, the material will pick up additional moisture and the texture will loosen.

The mixture created is called a 50/50 *admix*, even though there is twice as much cement as sand. That is so because the weight of the sand is one-half that of the cement, meaning the final mix is actually equal weights of each.

If the plaster in the pool or spa is colored, add the color powders slowly and keep mixing until you reach the approximate shade required.

The most important consideration in creating patch mixture is to consider the drying time and conditions. If the plaster dries too fast, it will shrink and crack. If the material is too thick, then it will not smooth out when you apply it and the result will be rough and unsightly. As noted previously, practice before going out to the job site.

The above recipe will make enough material to fill an 18-inch-diameter (45-centimeter) patch about ½ inch (13 millimeters) deep. Remember, once you start the job, it might not be practical to mix more material and if you do, it might not appear the same on the repaired area as the first batch. Therefore, always mix more than you need.

4. **Patch** Apply the patch with a trowel, using broad strokes, trying to fill the entire area on the first application. Try to fill so that the patch

material is higher than the surrounding plaster level, then remove the excess by scraping the trowel edge across the surface, making the patch clean, smooth, and level. Actually the best finishing tool for this smoothing is a straight piece of cardboard or plastic.

The other application method is underwater, where you work by hand with a fist-size ball of material and push it into the patch area with your fingers. Again, use more than you need so you can scrape off the excess, feathering the edges into the existing plaster for a smooth finish.

One last plaster patching technique is used to fill small surface cracks. Not all visible cracks allow water to penetrate the subsurface or cause delamination. Some are from too rapid drying or minor settling of the surrounding land or the spa itself. If there are no blisters, patch only the cracks.

The objective of patching small cracks is to slightly widen the crack so that it will accept patch material. You can use a small chisel to create a slight V shape along the crack. Follow the patching directions as outlined previously, either troweling on the patch material or rubbing it into the crack with your fingers and smoothing over the resulting repair with a straightedge.

Perhaps the most important aspect of any cosmetic patchwork is to have realistic expectations. Even the finest patchwork will be visible for a few weeks, simply because the new material is clean and the existing plaster has darkened with age. Colored plaster takes time to mottle and blend.

Coping and Tile (RATING: ADVANCED): When the ground around a spa shifts, which might be the result of settling or water leaking and causing erosion, the clues that this is happening might include popped tiles and coping stones. Coping stones are made of cast concrete, set edge to edge, so there is little room for expansion. A loose coping stone or one that has risen from the edge of the pool means there is some more extensive problem underneath. Rarely has a loose coping stone simply come free from the mortar in which it was originally set.

The objective, then, is to remove the loose coping, excavate the underlying deck, determine the cause, and relieve the pressures that created the problem so the same coping, or more, won't come loose again. Loose tiles are also an early sign of pressures behind the location

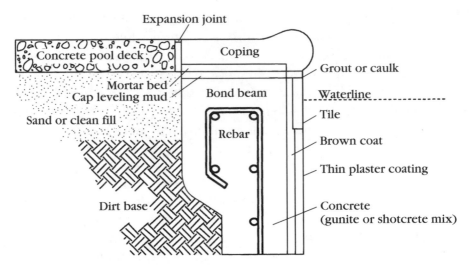

Expansion joint

Concrete pool deck

Coping

Grout or caulk

Mortar bed
Cap leveling mud

Bond beam

Waterline

Sand or clean fill

Tile

Rebar

Brown coat

Thin plaster coating

Dirt base

Concrete
(gunite or shotcrete mix)

FIGURE 9-10 Coping stones and tile line.

that need greater attention. The steps to diagnosing and repairing loose tile or coping stones are as follows (Fig. 9-10).

1. **Inspect** Lift off any loose coping stones and look for others by tapping each one with a hammer or the handle of a screwdriver. A hollow sound will quickly reveal other stones that might still be firmly in place, but that are not actually supported by anything.

   In many cases, you will need to remove the tile beneath the affected coping. It might have come loose already. Drain the spa to a level a few inches (or centimeters) lower than the bottom of the tile line. On larger jobs, it is easier to work inside the spa, so you will want to drain it completely.

2. **Cut** Remove the tile beneath the affected area by cutting through the grout and mortar that separate the bad section from the good. Also cut into the spa wall beneath the affected tiles. The cut should be about ½ inch (13 millimeters), made with the diamond blade on a handheld electric grinder. This cut prevents spreading cracks or chipping of the adjacent undamaged sections of plaster and tile.

3. **Chip** On small repair jobs where you have not completely drained the spa, you might want to float a piece of plywood under the

work area to catch as much debris as possible. Using a broad, flat chisel, chip the tile away to expose the mortar bed and/or spa wall beneath. If you are careful and have made adequate cuts, the tile might remain intact, but don't count on it. Make sure you have replacement tiles of the same design or that you can purchase an acceptable substitute. To make reassembly easier, mark the tile or lay it on the deck behind its location so you can return it to the same spot. When you remove each tile, place scratch marks (or use an indelible felt-tip marker) to note the exact positioning of each tile to make replacement easier.

4. **Open** Remove the stones. If the coping stones were not loose already, cut the grout joint between each one to make removal easier. Use a concrete saw, available at any tool rental store, with a 12-inch (30-centimeter) diamond blade. Cut along the mortar at least 4 inches (10 centimeters) deep to free each end of the stone. The back side of the stone should not be connected with mortar, but rather with a flexible expansion joint mastic or silicone that will not hinder removal.

    Standing on the deck, you should be able to grab the stones by the nose (the side facing the spa) and pull them free. If not, you might need to drive your chisel underneath the stone, which is now possible because you have removed the tile. As with the tile, mark the stones and lay them out on the deck so they can go back where you found them.

5. **Clean** Chisel and clean the underlying area as needed. Remove old expansion joint material, and examine the area between the pool deck and the bond beam. If coping and tiles have come loose, you will probably discover that there is no space between the two to allow for shifting and expansion. This is the cause of the problem, and it will cause more loose stones and tiles in the future unless the pressure is relieved and an expansion space provided.

6. **Joints** The objective is to create an expansion joint area about ½ inch (13 millimeters) wide and deep, enough to totally separate the deck from the spa wall. Chisel or cut away any material that is pressing against the spa wall. Never cut the spa wall or bond beam to create the expansion joint.

7. **Fill** When the expansion joint is complete, fill in any dirt that might have eroded away to complete the backfill area. If severe erosion has occurred, you might need to demolish a larger portion of the deck to expose the amount of backfill lost, replacing it before continuing. In such cases, work with a general construction contractor to rebuild the deck.

8. **Patch** Prepare the plaster patch material as outlined on p. 253. Some technicians add a latex additive to give the patching compound resiliency, but the best advice is to match the surrounding material. If the existing mortar is somewhat flexible, the new material should be as well. You can also use a premixed waterproof product such as Thoroseal, applying two thin coats before resetting the stones. The advantage of using a waterproof mortar is that you prevent water from weeping or leeching into the backfill again.

9. **Reset** Clean the stones of dirt and old, loose mortar. Apply a light coat of patch material to the underside of the stone, then sufficient patch material to the mortar bed to bring the stone up to its original level. You want to raise it slightly higher than the adjacent stones so that when it is pushed into place, it will settle down to the correct level. Tap the stones in place with a rubber mallet.

   Be careful not to allow the patch material to fall into the spa. If it does, be prepared to clean it up quickly so you don't create unsightly plaster chips in the spa.

10. **Retile** Prepare a brown coat of mortar to reset the tiles. The preparation of the bed is the most important step, because there might be high spots of old grout or mortar left after the original removal of the tile. Grind these down with a handheld electric grinder. It is better to grind too much, which can be filled with new mortar, than to leave high spots that prevent the tiles from reseating. Follow the marks you made to replace the tiles in the same locations.

    Thoroseal-type products work well and prevent water intrusion when the job is done. You might need to mix it with less water so it will hold the tiles on the vertical. Apply the tiles in the same manner as the coping stones.

11. **Grout** Regrout the stones and tiles. Grout can be premixed material purchased at the supply store or hardware store, or it can be mixed

by combining 1 part white cement with 2 parts sand. Use #60 silica sand unless the joints are over ½ inch (13 millimeters) wide (then use #30). Mix the grout to a loose enough texture to ensure it will completely fill the voids between the stones. Overfill, allow the mixture to set up slightly, then wipe away the excess to a smooth, level surface. Do the same with the tiles, carefully wiping all excess off the surface of the tile. A wet sponge wiped over the finished tile and coping will remove any grout film, which will otherwise leave a discoloration looking like paste wax.

Some spas use a colored grout on the coping stones or tile. Powdered dye is added to the grout and mixed until the color matches that of the existing installation. Remember, the finished job will not match until the new work has had a chance to weather and acquire the same shade as the surrounding work.

12. **Seal** Complete the expansion joint. Fill the joint with sand up to the last ½ inch (13 millimeters). Fill the last ½ inch with flexible mastic or silicone joint sealer, which can be poured as a liquid or injected like caulk. Follow the product label directions for application, especially concerning temperature and humidity ranges.

If you understand the basic underlying construction of the coping and tile area of the spa, you will be able to make these basic masonry repairs. Many will not be as complicated as described, requiring only rehanging a few tiles or resetting a single loose stone. If that single tile or stone keeps coming loose, however, or if more than one is loose, follow the procedures described earlier to determine the cause and effect a long-term repair.

Acrylic Spas (RATING: PRO):   Cracks in molded fiberglass or acrylic spas can be patched. To detect the leak, follow the previously described methods using evaporation, dye, or drain-down. Leaks in spas often occur where plumbing meets the spa shell. Shells often move away from their deck supports when the ground shifts and are especially susceptible to soil erosion problems. Since bather loads and displaced water are high in relation to the total size of a spa, water frequently washes away fill or base materials, allowing the weight of bathers to shift the spa and separate plumbing.

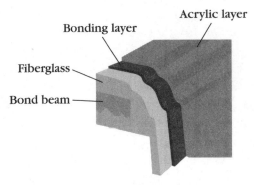

Acrylic layer

Bonding layer

Fiberglass

Bond beam

FIGURE 9-11 Acrylic/fiberglass spa construction.

Cracks or breaks in the acrylic can be repaired with materials available at the supply house. The technique is similar to patching small cracks in plaster, where you widen the crack to accommodate a fill material. The only difference with acrylic spa material is that the filler is not the final step; there is also a topcoat. (See Fig. 9-11.)

1. **Prepare** After identifying the hole or crack, drain the spa to a level below the work area and dry it thoroughly.

2. **Drill** Drill a small hole at each end of the crack to prevent it from spreading. The hole should be slightly wider than the crack, but not more than ¼ inch (6 millimeters).

3. **Widen** Using a file or a small blade on a handheld electric grinder, open the crack to a V shape wide enough to accommodate the fill material. Clean the crack with acetone to remove any dust or loose material.

4. **Patch** Patch the crack with acrylic filler, which is available in kits from the supply house. Modern filler is designed to expand and contract at the same rate as the spa shell, which is subject to extremes of temperature, chemicals, and bather load weight stresses. Follow the package instructions for mixing, curing, and drying times. Generally, acrylic patch sets up rapidly and dries in a few minutes, so be prepared to work quickly. As with plaster patching, make more material than you think you need so you don't run out in the middle of a job. Slightly overfill the crack, then wipe off the excess. Slight shrinking takes place during drying.

5. **Sand** When it is completely dry, sand off the excess down to the level of the surrounding surface. Avoid sanding too much of the surrounding surface. Because the acrylic finish is usually very high-gloss, imperfections in the sanding will stand out like the proverbial sore thumb. Take great care to smoothly sand the patchwork.

6. **Paint** The patch must be painted to match the color of the spa. Color powders or liquids are provided in the kit to mix with the base enamel paint, also provided. Mix very small amounts and

paint a test area, inside the skimmer, for example, until you achieve a good match. Remember that when the paint is wet, it appears somewhat darker than the final color.

Paint the patch, smoothing out the color and feathering it into the surrounding area. Spraying instead of brushing will achieve smoother results. In either case, remember that the match will not be exact.

7. **Seal** A clear topcoat seals the paint job and adds luster to the finish. The topcoat is also provided in the repair kit. Adding a little of the final color mix from the base coat to the topcoat will also help the finished repair blend better with the background material.

Do not use automotive body resins and Bondo-type products to repair spas. That was the only repair material available for many years, and experience taught that these products quickly shrink and leak when exposed to the harsh spa environment of heating, cooling, chemicals, and weight pressures.

Spa Remodeling

There are many ways to spruce up a tired-looking spa, including simple acid washing (for plaster surfaces) and painting (for plaster or acrylic surfaces).

FIBERGLASS COATINGS

RATING: PRO

Few gunite spas have an original interior surface of fiberglass because such applications are relatively new. The technology for surfacing gunite spas with fiberglass was imperfect 10 years ago, often leaving these spas with surfaces that were discolored, cracking, or delaminating.

In the past 10 years, however, numerous companies have improved the technology to the point where a fiberglass surface will give excellent results for many years. There are insufficient data available to evaluate if such coatings exceed the performance of plaster, but the obvious benefit of a fiberglass coating is that water and maintenance chemicals cannot corrode and destroy it like plaster.

Fiberglass coatings (Fig. 9-12) require the same surface preparation as replastering. The best mechanical bond is achieved by roughing up

A

B

C

FIGURE 9-12 (A) Fiberglass coatings for plaster spas. (B) Applying fiberglass coating over plaster. (C) Finished fiberglass coating in spa.

the surface prior to application. Fiberglass sheeting is laid on the spa interior and painted in place with chemical resins and fixatives.

Because fiberglass coatings and their applications are a unique technology, you should find a reliable contractor who specializes in such work.

INEXPENSIVE SPA FACE LIFTS: PAINT

RATING: PRO

The purpose of coatings of any kind over the gunite material of the shell is to prevent leaking, because concrete itself is porous. Modern spa paints are an attractive, inexpensive way to coat gunite, fiberglass, plaster, or any other interior surface with a smooth, colorful, water-proof coating.

Like fiberglass coatings, older formulations of paint were unreliable, often peeling and fading in 2 years or less. Paints were easily damaged by heat and harsh chemicals and were considered a temporary fix at best. Today, many paints will last up to 5 years, providing an inexpensive alternative to other coatings.

As with other coatings, the success of painting depends largely on the preparation and qualities of the subsurface being covered. If the paint is to cover old plaster, it won't keep delaminating material from peeling from the shell surface. In other words, before painting, evaluate the surface being covered. If it won't last for three to five years more, then there is no value in painting over it. It is time to replaster or remove the old plaster before painting.

It is also important to have realistic expectations about paint as a surface covering. Irregularities in the subsurface will show through the paint and will still be felt underfoot in shallow areas. Colors might be vivid initially, but will fade throughout the life of the paint—especially the brighter the original color. Finally, paint will not last much longer than 3 to 5 years and might begin to peel or dissolve prior to that.

Spa and pool paint is manufactured in three types. Chlorinated rubber is designed to be flexible when applied, therefore it is durable and appropriate for rough, previously unfinished surfaces, acting as a stretch-rubber cover. Solvent-based paints are also good on rough surfaces because they are designed for thinning with additional solvent, making them easy to spray on. Water-based epoxies are used for painting fiberglass pools or spas and are more resistant to chemical variations and temperature extremes.

Generally, cover old paint with new paint of the same type. You can use water-based epoxies to cover chlorinated rubber, but not the other way round. If you're unsure, sand off the old paint or consult the manufacturer (both are good ideas anyway). Another simple test method is to rub a little of the new paint over the old painted surface. If the old material dissolves or bleeds, don't use the new product without removing the old.

Also, dark colors will cover more evenly and hold up longer than lighter colors, with the worst being blue.

If you are prepared to proceed, follow the paint manufacturer's label directions and these guidelines.

1. **Prepare** Prepare the surface. Empty, clean, and remove any loose materials from the spa. Sand any loose plaster or high spots. If there are blisters, follow the plaster repair procedures to prepare a level surface. Remember, paint can't replaster or level out a spa surface. If the surface is fiberglass, rough-sand the entire area to be painted for a good mechanical bond.

   Because sanding, especially if it requires removing old paint, requires enormous amounts of time and sanding disks (for your electric sander), test a few square feet (1 square meter) before estimating the cost of the job for your customer. I have spent more on the preparation, in time and materials, than I quoted to do the entire job, simply because I thought it would take less to get it done.

2. **Clean** Clean the finished preparation job. If you have spent many hours creating a level, paintable surface, don't waste that effort by leaving dust and dirt on it that will prevent the paint from adhering. Wash down and scrub the surface with TSP [1 cup per 1 gallon (250 milliliters per 4 liters) of water and 1 ounce (29 milliliters) of tile soap to cut any oils on the surface]. Make sure that you are using pure TSP, not products labeled as such but containing additives that might leave residues.

3. **Acid-Wash** Acid-wash the entire area to be painted, following the acid-wash directions detailed later in this chapter. When acid-washing for a paint job preparation, the idea is to etch the surface to create a roughness that will help the paint adhere. In regular acid-washing, the idea is to clean the plaster but not to etch. In other words, for a paint job use a stronger acid mixture and leave it on the surface longer.

4. **Clean** Scrub and rinse the entire surface with TSP again to remove any trace of acid, this time without any soap. Rinse and scrub again with clean water. This is not the time to skimp on labor, because any remaining acid or dirt will prevent the paint from adhering. The entire spa might be flawlessly painted, but you will only notice the 3-inch blister or the small crack in the final paint job.

5. **Dry** Allow the surface to dry thoroughly, and don't plan to paint on foggy or humid days when the moisture content in the air is high. Similarly, avoid working during hot afternoons when the surface to be painted is hot. Either condition will keep the paint from bonding to the surface or will cause blisters. As with all painting, follow the product directions. Mix the paint, enough to do the entire first coat. Mixing might mean simply stirring, or it might mean adding catalysts according to the manufacturer's directions.

6. **Paint** Use good-quality tools to apply the paint. Thin, cheap rollers will buckle under the pressure needed to force the paint into the

---

TRICKS OF THE TRADE: SPA PAINTING

Most good spa paint jobs are 90% preparation and 10% execution. Here are tips to make the job easier and the results more satisfactory:

- A smooth (patched if necessary) surface yields the best paint job, so after cleaning the surface to be painted, identify areas that need repair by running a rigid putty knife over the suspect areas such as the underside of fittings. Many paint manufacturers also make patch kits that are specially designed to facilitate painting, so use them.

- If the surface to be painted is very rough, use epoxy paint. You can apply several coats, building one on the other, creating a nice, smooth shell.

- Remember that the surface and air temperatures may differ, so don't start painting if it's approaching the warmest part of the day and the spa has been in the sun for awhile, even if the air temperature still feels cool.

- If you are painting a surface that has once been painted and you don't know what type of paint was originally used, try this test. Dab a little chlorinated rubber paint on top of the old surface and wait about a minute. Rub the area with a rag. If the old paint rubs off with the new, the original paint was also chlorinated rubber. If not, it was probably epoxy-based paint.

- Indoor spas should be painted only with water-based paints since they contain no noxious solvents which can create fumes. Allow 2 to 4 more days for proper drying before refilling, and use fans throughout to facilitate the process.

- Always read the labels on the paint cans thoroughly—you'll be amazed what you can learn and the mistakes you'll avoid!

- Stir the paint often, especially if it's a colored paint, to ensure even appearance.

crevices of the roughed-up surface, especially on the first coat. A 9-inch-wide, ½-inch-nap (23-centimeter-wide, 13-millimeter-nap) lambskin roller on a sturdy frame works well. First, cut in (detail paint that can't be done with a roller without overpainting) the tile, drain, light, and other detail areas with a good bristle brush, then roll on the paint. Apply a thin but even coat. The first coat will use 2 to 3 times as much paint as subsequent coats, except when applying over fiberglass.

7. **Add Topcoat** After waiting the manufacturer's recommended time between coats, apply a second and, if necessary, a third coat. On all coats, but especially these finish coats, paint the entire surface at one time. If you stop and restart, no matter how careful you are, the dividing line will still be visible with slightly different tones of color. On the last coat, sprinkle some clean sand over the wet paint on the steps and in the shallow areas to aid traction for bathers.

8. **Cure** Let the paint cure as recommended by the manufacturer, then refill the spa. Avoid using the heater for a few more days, just to be sure the paint has thoroughly cured.

As with any repair procedure, but especially with paints, work carefully and clean, taking care to spread dropcloths on decks and over deck furniture. Clean up any spills promptly and clean tools according to the manufacturer's suggestions, preferably back at your workshop. Remember, the customer will only see the blemishes of the job, no matter how few and no matter how good the rest of the job looks.

Maintaining a Spa with a Painted Surface: Maintenance of painted surfaces is the same as that of other spa coatings, but since there is no plaster to raise the pH, pay special attention if you need to add acid. Sometimes the painted surface appears to have a chalky coating, which is easily removed when brushed. This is not the paint "curing," as some will tell you. Chalking usually occurs when the total alkalinity has dropped too low and some minerals come out of solution and deposit on the walls of the spa, essentially scale, which is common to all spas (but usually appears only on tiles). To fix the problem, adjust water balance as previously described and use a clarifier or flocculating agent to help filter out the unwanted mineral. Brush often for several days and keep the filter running 24 hours per day to finish it off.

ACID-WASHING

RATING: ADVANCED

Perhaps the simplest remodeling technique for a plaster surface is an acid wash to clean the plaster surfaces of the spa. As discussed earlier, it is necessary to drain the spa from time to time, to remove scale-causing minerals and chemicals. This is a good time to acid-wash the spa, to brighten the appearance and remove any deposits that might have built up.

As with other cosmetic procedures, the key to success is realistic expectations. If there is severe etching and staining, an acid wash will not smooth out rough plaster or remove every discoloration. It will make the spa look better, but it might still feel rough underfoot and show signs of metal deposits and/or scale from years of abuse. Especially with colored plaster, you must be careful not to remove so much scale and/or stain that the acid also removes substantial amounts of plaster, leaving pitting or gunite showing through.

Before acid-washing, it is important to know how old the plaster is and if it has been acid-washed before. If the plaster is nearing the end of its useful life (10 to 15 years), there might not be enough material to wash without stripping the surface down to the gunite. Similarly, if the spa has been acid-washed 2 or 3 times already, it might be time to consider replastering, painting, or some other recoating.

In each case, once you drain the spa, test a few areas before continuing the job. If your tests reveal the plaster is thin or already badly etched, discuss expectations and alternatives with the owner.

1. **Shut Down** Turn off the circulation equipment and in-water lighting at the circuit breakers.

2. **Drain** Drain the spa, following the techniques described previously.

   Position your pump near the main drain, unless you know the lowest point of the spa to be elsewhere. If you're down to the last 20 to 30 gallons (75 to 113 liters), you might want to leave it in there as a dilution pond for the acid you will be using. If you use the technique of directing the wastewater to a lawn or garden for irrigation, remember to redirect the discharge hoses before you actually begin the acid wash. Direct the hoses into a deck drain or run them out to the street.

3. **Set Up** When the spa is drained and you arrive to perform the acid wash, bring your acid and tools into the empty spa. Set up your equipment on the stairs. Hose off the deck as soon as you arrive and continue to keep it wet during your work time to prevent inadvertent staining or spills. While we're on the subject, once you begin, wash your boots off each time you exit the spa. They carry acid as well.

4. **Prepare** Remove the light fixture and any other loose or removable hardware such as ladders, rails, or metallic return outlet nozzles. Remove the main drain cover. Clean out the main drain and light niche. Rinse (and scrub if needed) any other organic material (oils, leaves, and dirt) from the plaster, because the acid will not dissolve these and therefore will not clean the plaster beneath.

5. **Look** Inspect the spa for plaster blisters, as described previously. Make any necessary repairs or patches before acid-washing, so the subsequent acid wash will clean all surfaces to approximately the same color, helping any patches to blend.

6. **Take Precautions** Before you start the acid wash, put on old clothes (a few small acid splashes will put holes in cloth), rubber boots, gloves, and a respirator (rated at least R25). The boots provide skin protection and insulation against shock if the pump is shorted out in the wastewater around the main drain. I have done acid washes in bare feet with no protection and have known technicians who have not used protection during 30 years of performing acid washes, but the cumulative effect of inhaling acid fumes, exposure to raw acid on the skin, and the effect of fumes on the eyes can't be measured. Moreover, I have experienced some really scary incidents when acid washing in varying wind conditions, where a cloud of acid fumes suddenly changed direction and enveloped me, choking and burning with sobering effect. Remember, too, that if any chlorine is still present on the plaster, when the acid hits it, it will create lethal chlorine gas. In short, believe me when I say **always** use a respirator, gloves, and boots.

7. **Use the Water Supply** Finally, prepare a garden hose that can reach all parts of the spa and keep the water running. Although leaving

the hose on will waste a bit of water, it will also keep the floor of the spa wet in the area you are working, neutralizing the acid as it runs toward the main drain and the pump. In this way you can control the area the acid covers and the length of acid contact.

8. **Test** Take a garden-type sprinkling can, add a squirt of tile soap, and fill it two-thirds with water, one-third with acid (always adding the acid to the water, not the other way round). The soap will help the mixture cling to the walls of the spa and will cut the fumes somewhat. If you add the soap first, you won't need to stir or mix the solution. *The effectiveness of acid washing is a function of the strength of the mixture and the length of time it contacts the plaster.* To determine what the particular job needs, test the solution on a small area by pouring a stream from the sprinkling can. Allow it to sit on the plaster for about 30 seconds while scrubbing the area with a stiff-bristle broom or brush; then wash it off with clean water from the hose.

If the stains haven't disappeared, you might need to leave the mixture in contact longer, strengthen the mixture, or go over the area more than once. Trial and error is the only way to become familiar with these variables, especially since you might encounter several different problems within each job. The one-third acid mix might work fine for most of the spa, but a few areas might need a 50% solution or even straight acid to budge tough stains. I have had to apply pure acid and leave it for up to 45 seconds on some stains to get them clean, so be prepared to spend some time experimenting on each spa to determine the needs of the job.

One word of caution, however: you don't want to cure the illness, but kill the patient. It's not worth removing every last bit of discoloration or scale, just to end up with extremely rough surfaces or plaster so thin that you can see the gunite showing through. Therefore, the other reason for conducting some trial spots and solutions is to determine the effect on the plaster as well as the stains.

If you find stronger solutions and longer contact times having little effect, you might have an older plaster mix composed of a silica sand base, rather than the more erosive modern mixtures of crushed marble (marcite) and calcium chloride. Don't be afraid to try stronger mixes or pure acid if that's what it takes to achieve good results.

As a rule of thumb, a typical 1000-gallon spa usually requires about 2 gallons of acid (or about 4 liters of acid for every 4000 liters of water). You might not use it all, but it is better to be prepared.

9. **Acid Wash** Once you have determined the acid strength and contact time for the job, continue around the spa with the sprinkler can of acid solution, the scrub broom, and hose for rinsing. Keep rinse water flowing on the spa bottom when you are not actually rinsing an area that has been acid-washed to neutralize acid on the spa bottom. Keep the pump operating to remove the waste from the hopper (the low point in the spa around the main drain). Even though the acid solution is somewhat neutralized by the plaster it is dissolving, it is still very potent. If you allow it to run undiluted along the bottom as you work the entire spa, when you finish, you will have acid-washed the bottom and the hopper several times.

If your area establishes a minimum allowable pH for waste-water, you might have to toss 1 cup (250 milliliters) of soda ash into the hopper and mix it with the broom from time to time to bring up the pH of the discharge you are pumping. Some jurisdictions prohibit a discharge with a pH less than 5.0, and it can take up to 8 pounds (3.6 kilograms) of soda ash to neutralize 1 gallon (3.8 liters) of acid, so make sure you have adequate soda ash on hand when you do the work (soda ash is very inexpensive).

The purpose of scrubbing with a broom is to force the solution into the pores of the plaster for thorough contact. All the scrubbing in the world won't actually remove stains, so let the acid do the work by making sure it is in complete contact with every part of the plaster. The most effective way to work is to have one person apply the acid and another person scrub. Either can rinse after they finish. If you must work alone, use a more diluted concentration, allowing you to leave it in contact with the plaster longer and giving you ample time to apply, scrub, and rinse.

10. **Observe Safety** For safety reasons it is better to work with a helper. Acid-washing not only cleans the plaster, but also leaves it smoother than before, as it removes scale and rough spots in the plaster. Since the smooth surfaces are also wet, it is easy to

slip and fall. For this reason, I try to work around the sides of the spa, then the bottom.

11. **Clean Tile** Take advantage of the spa being empty and clean the tiles. Under normal conditions, with water sloshing along the tile line, it is often difficult to remove all the oils and scale on the tile. With the spa empty, scrub the tiles thoroughly, using tile soap, pumice, or the barbecue brush and, while you have it, some acid solution. If the tiles are particularly dirty, you might want to clean them before the acid wash so that any runoff washes over the plaster before it is cleaned. Use a pumice stone or block to scrub out any especially stubborn stains in the plaster.

12. **Rinse** Finally, thoroughly rinse the spa and lightly scrub any remaining acid or soap residue, so that when you refill the spa, it will not have negative effects on the pH. You might want to go over the steps and shallow area, where feet contact the surface, with an electric sander and some fine to medium wet/dry sandpaper (80- to 100-grit) to smooth out any rough spots.

13. **Clean Up** Give the light fixtures and other hardware a good cleaning before reinstallation. Metal fixtures can be lightly acid-washed, and the plastic ones can be scrubbed with cleanser and a wire brush. Use a cup to remove any acidic water from the main drain before reinstalling the cover. Reassemble what you took apart and remove your equipment from the spa. To avoid trailing acidic mixtures across spa decks and through access areas. I rinse and then put the pump, hoses, sprinkling can, and empty acid bottles in big plastic garbage bags before removing them from the pool. Another option is to put everything in a large plastic garbage can as a way to collect and carry everything back to the truck without spilling. Even a few drops spilled on walkways and steps will be noticeable, and you might find yourself acid-washing all the concrete around the house for free to erase the evidence!

14. **Fill Up** Fill the spa. Before you restart the circulation, make sure the filter and strainer basket are clean so that dirt doesn't flush into the spa and onto the clean plaster.

15. **Start Up** Brush the spa while circulating the water for a few minutes, then check the chemicals and balance (as described in

the section on restarting after a new plaster job). If the pH or total alkalinity is extremely low, turn off the circulation, balance the chemistry, and brush to circulate your chemicals before restarting. If you didn't do a good job removing excess acid, resulting in corrosive water, you might strip copper or other metals from the spa equipment, staining the plaster all over again.

REMODELING THE DECK

RATING: PRO

One of the more dramatic improvements that can be made to an old spa is not necessarily in the spa itself, but rather a remodeling of the deck. The variety and texture of deck surfaces are limited only by your imagination.

Since deck work, including brick, wood, and the "new" surfaces discussed here, requires special training and talent to install, we will not go into detail on this topic. The point of presenting some basic information is to assist in making a choice of new deck materials. Here's a sampling of choices that are readily available today:

- Traditional cement and sand combinations, but sprayed or brushed with colors to create a more natural rock appearance.
- Cast-in-place colored concrete that is stamped by special forms to create the appearance of wood, brick, stone, or flagstone.
- Epoxy stone coverings, composed of small colored gravel held together by a clear, waterproof adhesive bonding material.
- Deck paints with nonskid texture incorporated.
- Natural cobblestone, flagstone, brick, marble, or sandstone. Real stone offers an appearance and texture that are hard to equal, but is more costly to install.
- Wood is a low-cost alternative that provides natural beauty which changes in color as it weathers (Fig. 9-13). Wood decks must be sanded and retreated with waterproof coatings periodically to prevent cracks and splintering.

Many of these treatments are continued from the deck to the coping to make a seamless appearance around the spa. This concept is espe-

FIGURE 9-13 Modern spa deck. *Gordon & Grant Hot Tubs.*

cially striking with natural materials (or those that appear natural), creating the appearance of a natural pond among the rocks.

Hot Tubs

Wooden hot tubs (Fig. 9-14A and B) enjoyed great popularity in the 1970s and then largely disappeared during the 1980s. In recent years they have made a comeback as people seek simple, natural surroundings and as they discover the economy of wooden spas. Figure 9-14C shows a cross-section of a typical wooden hot tub. Made from any dense hardwood that is resistant to moisture and decomposition (usually redwood), hot tubs are assembled from dozens of vertical boards called *staves*. The edges of each stave, which look like a standard 2- by 4-inch (5- by 10-centimeter) plank, are beveled (angled) slightly so that when the staves are assembled edge to edge, the result creates a circle. The bottom of each stave is carved with a groove, called the *croze*, to accommodate the width of the floorboards.

The floor is composed of similar horizontal boards, but with straight edges. One edge of each floorboard has two or more holes, and the other edge has corresponding pegs. This aids in assembly, keeping the floorboards properly aligned, although the pegs provide no actual structural support. The ends of each floorboard are curved slightly, and the center floorboard is the longest, with each succeeding floorboard on either side

A

FIGURE 9-14 **(A, B) Redwood hot tub. (C) Cutaway view of wooden hot tub.** *A, B: Gordon & Grant Hot Tubs.*

being slightly shorter so the ultimate result is a circular floor. The floor is supported by 4- by 4-inch (10- by 10-centimeter) joists. The tub does not actually rest on the staves; rather the floor of the tub rests on the joists.

The tub is assembled in the ancient tradition of cooperage (barrel making), with the staves being held together by curved steel bands that are tightened by passing each end through a hoop lug (a hub connector fitting) and secured by nuts screwed down on each threaded end. The tub is made watertight when water is added, swelling the wood and sealing the seams between the staves. Wood seats are added by fastening bench supports with stainless steel screws.

Wood is specially selected and prepared for hot tub construction. Smooth, even, vertical grain is preferred, the best being taken from the

B

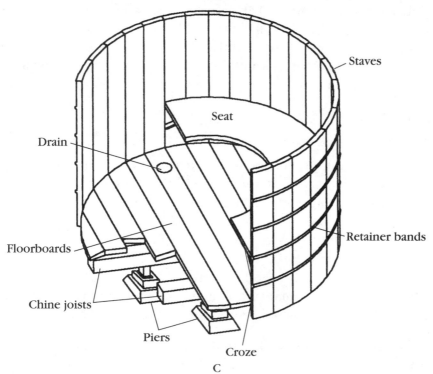

C

FIGURE 9-14 (*Continued*)

heart of the tree where the grain is best and the most tannin is located. Tannic acid makes the wood more impervious to the decomposition effects of heat, water, and chemicals. The wood is then kiln-dried so it is devoid of almost all moisture. Thus, when the finished tub is complete, the wood quickly absorbs moisture and swells for tighter seam sealing.

The woods used for hot tubs differ in various parts of the world depending on their availability and price. It is generally accepted that California redwood is the best hot tub material, but teak, cedar, mahogany, certain oaks, and several exotic rainforest hardwoods are also used.

Hot tubs are made in various sizes—the most common unit is 6 feet (2 meters) in diameter by 4 feet (1.2 meters) deep. I have assembled and installed tubs up to 10 feet (3 meters) in diameter by 4 feet (1.2 meters) deep with 20 separate jets.

One advantage of using wood for a spa is its natural insulation property. Simplicity and blending with the environment of modern gardens and homes are the other. The disadvantages are lack of comfort (you can't mold lounges in a hot tub as you do in a plastic spa), the wood maintenance requirements, leaking problems, and bacterial questions. Although a properly maintained wooden hot tub is as clean and safe as a gunite or plastic spa, many health and safety codes prohibit wood for commercial or public use because bacteria can be held in organic materials such as wood and transmitted to others.

Installation

RATING: PRO

Each manufacturer provides assembly instructions with its precut tub components. The manufacturer I use will assemble the tub for me for an extra $100—well worth it to avoid the job myself. Unfortunately, most tubs are too big and heavy to carry through the narrow passageways alongside most homes, so assembly in the homeowner's backyard is usually required.

Installation of the hot tub proceeds along the same lines and requires similar equipment and plumbing as a plastic spa. The following guidelines detail the few procedures that are unique to hot tubs.

1. **Foundation** After you work with the homeowner and/or general contractor to determine the best location for the hot tub, prepare a stable

foundation. Hot tub wood needs to breathe, to be surrounded by air. I have been called to installations that were backfilled with sand or dirt, usually for leak problems, and have always discovered extensive wood rot. The tub requires lateral support (side to side), and the most important aspect of the installation is the foundation. Because the weight of the tub, water, and bathers is enormous, do not mount a hot tub on a patio or wooden deck that has not been reinforced.

If you are not using a general contractor, prepare your own foundation. Prepare a level, well-compacted area of ground. Create a form for the concrete using 2- by 4-inch (5- by 10-centimeter) lumber as shown in Fig. 9-15. Excavate the area inside the form to a depth of at least 2 inches (50 centimeters). The completed foundation sits in the ground, rather than on top of it, and will be resistant to shifting or earth movement. Lay ½-inch (13-millimeter) rebar in a crossed pattern as shown, creating 1-foot (30-centimeter) squares. Tie the

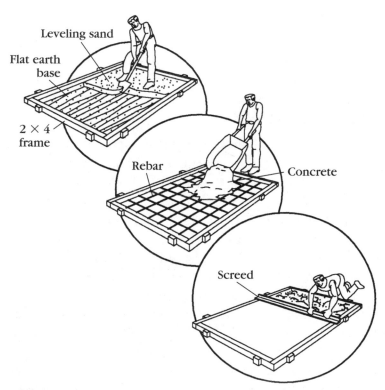

FIGURE 9-15 Building a concrete foundation.

intersections with tie wire to secure the grid. Slide small blocks of wood or brick under the resulting grid to elevate it about 2 inches (50 millimeters) off the ground. Prepare a concrete mix, available at the hardware store, and mix as directed with sand. Pour the form full of concrete, forcing the mixture under and around the rebar grid. Make a screed of a 2 by 4 inches (5 by 10 centimeters) and level the concrete as shown. Allow the foundation to cure for at least 4 days before assembling the tub on it.

An alternate foundation can be made of precast concrete piers, available at masonry supply yards. Set the piers 18 to 24 inches (45 to 60 centimeters) apart in lines so that the tub's joists will rest on the piers. Pier foundations can be used on any solid ground that is not subject to erosion or further compaction.

2. **Plumbing** (Fig. 9-16) Because air must be left around the tub, you can usually set it in place before plumbing it and still have enough room to work. If working in tight quarters, preplumb the tub, as described in the section on spa plumbing. In either case, work with the homeowner to determine the location and height of the jets. Drill holes in the wood, taking care not to cut the retainer bands on the outside of the tub. Jet fittings are installed as described previously, but be sure to use jet throats designed for the thicker walls of a wooden tub (the throat is longer; the jet body is the same). Plumb the tub as outlined previously.

3. **Air Bubble Rings** If an air bubble ring is desired, you must create it from flex PVC. Install an additional main drain-type fitting near the bottom of the tub. The inside of the fitting is threaded for 1½-inch (40-millimeter) plumbing. The supply house will have a threaded T fitting that screws into this extra drain fitting, stubbing out to the left and right with ribbed nipples. Lay a circle of 1½-inch (40-millimeter) flex PVC around the edges of the floor of the tub. The ends of the pipe will fit over the ribbed nipples, creating a ring. Using the guide in the blower section, drill the number and size of holes into the top of the ring. The pipe from the blower to the tub can be glued into the back side of the drain fitting. When air is forced into the ring, the fitting will hold down one side but the other will float. Secure the ring to the floor with a stainless steel hook and a strap made from plastic tubing or some other nonmetallic

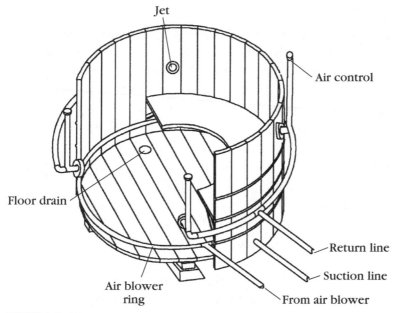

Jet

Air control

Floor drain

Return line

Suction line

Air blower
ring

From air blower

FIGURE 9-16 Hot tub plumbing.

material. Some air rings in hot tubs are installed under the seats,
toward the front edge (the edge at the middle of the tub). Again,
the fitting will hold one side, but the ring needs to be secured to the
bottom of the seat in at least one other place.

4. **Additional Fittings** Most hot tubs do not use a skimmer; however, if
   that is desired, install it as described previously. The other plumbing
   connections to the equipment will also be the same as with plastic
   spas. Since hot tubs require frequent draining, many technicians
   drill a hole in the floor of the tub and install a drain that leads
   through a 1-inch (25-millimeter) pipe to a hose bib outside the tub.
   This allows you to connect a garden hose so you can direct the
   discharge from the tub away from its base.

   Figure 9-16 shows an air blower line that serves an air ring. As
   with plastic spas, it can be plumbed with a three-port valve to alternate between the ring and the jets, or a second blower can be
   installed. Figure 9-16 also shows jets that are aerated from the
   atmosphere, but as described previously, the hot tub can be fitted
   with jets that are charged from a blower.

Break-in

A plastic spa can be filled with water and used immediately, but a wooden hot tub requires break-in. When you have completed the installation, fill the tub and keep the water running. Water will run through the seams for as long as 24 hours while the wood absorbs water and swells, sealing the seams. If you fill the tub only partially, the seams will close only up to that level, so keeping the tub full is important. Allowing the water to overflow will also moisten the outside of the wood, speeding absorption and swelling. Some tubs will seal in as little as 3 or 4 hours. In any case, keep the water supply flowing until the tub is no longer leaking.

As soon as there is enough water holding in the tub to start circulation, begin to run the pump. Turn the heater on as well, because that will speed the swelling of the wood and the leaching portion of the break-in. For several days, tannic acid and natural wood oils will leach from the timbers, turning the water coffee-colored. The best wood is taken from the heart of the tree where these acids and oils are most abundant, so the time it takes to leach them from the wood is an indication of the quality of the timber in the tub. I have installed some hot tubs that barely discolored the water, indicating the wood contained little protective tannic acid or oils.

Continue to run the circulation 24 hours per day for 2 days. Brush the inside of the tub vigorously once or twice each day to scrub out as much of the acid and oil as possible. There is no way to filter or chemically remove this discoloration, so drain and refill the tub at the end of the second day. It might take two or three such refills before all the discoloration disappears from the tub. Some technicians will tell you various methods to speed this process, including bleaching and soda ash preparations, but believe me, I have tried them all and the only thing that works is patience.

Although you are advised not to use the tub during the break-in, some have done so anyway without suffering any ill effects from the discolored water. It doesn't stain and apparently has no detrimental health effect on the skin, but I still ask the customer to stay out of the tub until the water is cleared. Each time you drain the tub, clean the filter and drain as much water from the circulation system as possible to remove all the affected water. When the water is clear, add sanitizer and operate the tub normally.

If leaks persist in a particular seam or part of a seam, take note of the rate of the leak. If it seems to be leaking a little less each day, it might be an imperfection in the wood and will seal in a few days. If the rate of the leak continues unabated, follow the leak repair procedures outlined in the next section.

Repairs

RATING: ADVANCED

Hot tubs are more prone to leaks than other types of vessels, simply because there are so many more places for leaks to develop. In addition to plumbing and through-wall fittings, hot tubs have dozens of vertical seams and the entire length of the croze seam where leaks can develop. In extreme cases, wooden tubs can also rot through the entire thickness of the wood itself.

LEAKS IN THE STAVE SEAMS

If leaks develop between the staves, it could mean that the bands need to be tightened. Rarely does a seam leak along its entire vertical length; rather a section of several inches to 1 foot (up to 30 centimeters) will leak along one seam. Tighten the band closest to the level of the leak. If the wood has dried out, fill the tub and allow the wood to reexpand.

If tightening the bands is unsuccessful, the seams can be caulked. Some technicians dry out the wood and apply silicone sealant; others use sealants designed for use on wet surfaces. These measures will work temporarily, but the wood and sealant expand and contract at different rates. This means that in a short time, the repaired area will leak again.

For centuries, builders of wooden ships have caulked seams with lamb's wool. Because it is organic, it will expand and contract very similarly to wood. Wool contains natural resins and oils, and it will remain impervious to decay for several years. Some technicians recommend twine or other organic materials, but lamb's wool will last the longest.

The best source of lamb's wool is the local grocery store. Mop heads are often made from lamb's wool (be sure you are not buying a synthetic imitation) and provide strands of wool about the right thickness when unbraided. Knitting yarn of pure wool will also work. Repair is a simple matter of pulling out strands of wool from your source material that are thicker than the seam being sealed and forcing the strands into the seam with a putty knife. Work from the inside of the tub with the water drained below the work area to make access easier. Don't allow the wood to dry out, however, because that might create even more leaks when the wood reswells.

When you have completed the caulking, refill the tub and heat the water. It might take a day for the wool to swell and seal the seam. During this process, water leaks out, pulling the wool deeper into the seam. If after 2 or 3 days the seam is still leaking, try the repair again.

Another method of sealing seam leaks is to add sawdust to the water, hoping that as the water leaks out, the seam will plug up with wet sawdust. This method only works 1 time out of 10 and is a very temporary fix. I have never heard of such repairs lasting for more than a few months. A similar concept is a product called Tub O'Gold, which is a chemical compound that is added to the water and allowed to leak through the seams. The compound is supposed to clog the leak, filling it with a white, pastelike material. Again, I have found this product to work once in awhile, but the results are not long-term. Moreover, this product requires extensive brushing, vacuuming, and filter cleaning to remove the excess dough that is created, which never seems to completely clear from the tub. After many experiences with such quick or easy fixes, I would recommend disassembly and reassembly—realignment of the staves—as a way to repair leaks before I would mess with sawdust or Tub O' Gold.

LEAKS IN THE CROZE

The seams between the floorboards rarely leak, although if they do you can caulk them as described previously. Leaks are more likely to be found in the croze. Leaking in the croze can be caused by pressure or wood decomposition. Figure 9-17 shows how a deck can apply pressure to the top of the staves, kicking them outward at the base near the croze and causing separation. The solution is to relieve the pressure and tap the base of the staves back in with a rubber mallet. A similar pressure leak can be caused if a band is tightened too much just above the level of the croze. In each of these cases, loosen and reposition the lowest band on the tub so that it wraps around the floor and croze joint. Tighten the band until the staves are replaced securely to the floor.

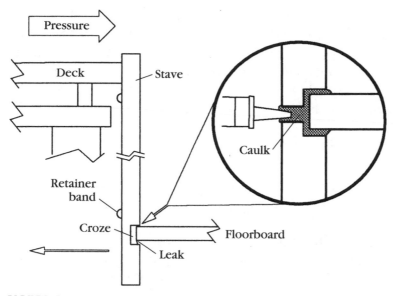

FIGURE 9-17 Hot tub repair.

Pressure leaks are also caused if the bottoms of the staves are resting on the deck, supporting the weight of the tub. Remember, the joists are designed to support the weight of the tub through the floor, not the staves. If the tub has settled for some reason, you can raise it and reposition the joists or piers by emptying the tub and using a car jack to raise it off of the deck. To raise it, apply the jack to a band, not a stave, or you will create more pressure damage. Use extreme caution when jacking a hot tub off the ground. The forces are extremely great, and the jack head is in contact with a small band of metal (which can snap if it is rusted or cracked); and if the tub shifts suddenly, the jack can snap out of place with deadly force. I have jacked many tubs to examine the underside or reset the supports, and I have always succeeded by working slowly and with helpers to steady the tub. You also don't want to take any chances when the tub is raised on a jack by reaching under the tub. Use boards or a telepole to push materials out from under the tub or maneuver new support pieces into place. A $20 telepole is easier to replace than an arm.

If the wood has deteriorated, the croze can be caulked with silicone sealant. Because it is virtually impossible to insert organic materials into the croze, this is the only time when use of a sealant is recommended. As shown in Fig. 9-17, drill a hole in the stave directly behind the floorboard in the vicinity of the leak. Fill the voids with caulk. You might need to drill holes in several staves to stop all the leaks. The hole should be no larger than necessary to accommodate the tip of the caulk tube. You can plug the hole with a short length of wooden dowel of the same diameter as the hole or with more caulk.

LEAKS IN THE FITTINGS

As with other types of spas, hot tubs can leak around jet, drain, or skimmer fittings. Wood is more likely to leak in these areas because the wood expands and contracts with temperature extremes, while plastic does not.

Always try to tighten fittings first. A loose fitting is usually the source of the leak. If that fails, follow the techniques already described. As wood ages and deteriorates, the hole around the fitting might enlarge or rot. In this case you might want to fiberglass (as described later) over the face of the fitting where it meets the wood.

LEAKS FROM WOOD DECOMPOSITION

As the tub ages, the wood will begin to decompose. This can be accelerated by poor water chemistry maintenance, but the process is inevitable.

I have added another 2 or 3 years to the life of a decaying wooden hot tub by fiberglassing the interior. The entire interior of the tub can be fiberglassed, or you can do just the croze joint or specific staves that are rotting or leaking. The steps are as follows.

1. **Drain the Tub** Allow it to dry for several days so that there is no surface moisture. If the wood is still moist deep inside, it can dry out by breathing through the exterior, because only the interior surface will be fiberglassed.

2. **Sand** Use a rough-grain sandpaper (40-grit) to strip off any decomposing wood and expose the best of the hardwood below. The roughness will also aid in creating a mechanical bond for the fiberglass. Wipe out the sanding dust completely, or wash the tub and allow it to dry again.

3. **Patch** Fiberglass patch kits are sold in most pool supply houses, all boat supply stores, and in many hardware stores. Fiberglass is actually

a collective term applied to two distinct parts of a process. Fiberglass itself is a cloth woven from yarns of various twist and ply construction into a variety of sheets, strips, tapes, or chopped loose fibers. The fiberglass provides the strength, but by itself is not stiff and will not adhere to a surface. Liquid resin is the other half of the fiberglass process. Resin is a clear or slightly golden liquid that is poured or brushed over the fiberglass fabric. The resin dries hard, providing the bond and stiffener, while the fabric provides the strength.

Unless large areas of the hot tub are deteriorated and the structural integrity of an area is in doubt, fiberglass cloth is not needed for this repair. The wood takes the place of the fiberglass cloth, providing the framework onto which the resin is applied.

Follow the product preparation directions on the patch kit. Generally, after sanding and cleaning the surface to be fiberglassed, you will mix a two-part resin to paint the surface to be repaired. One part of the resin mixture is resin, while the other part is a catalyst that starts a chemical reaction which leads to hardening. The two cannot be premixed, or the product will set up inside the can.

After mixing the resin components, liberally brush the mixture on the area to be patched. As the mixture thickens, it will adhere better to vertical surfaces, so you might want to wait a few minutes before application. In any case, most mixtures will set up within 10 to 15 minutes, so be ready to work quickly.

If there is structural support needed, apply a thin coat of resin, then lay up fiberglass sheets or strips. Allow 24 hours' drying time (or as recommended on the particular product you are using), then apply a second coat of resin. Sand lightly between coats to ensure good bonding of each coat. Always mix more resin than you think you will need, to avoid running out in the middle of the job. The first coat will require about twice as much as successive coats because the bare wood will absorb a great deal of resin.

When you fiberglass an exposed product such as a surfboard or a piece of furniture, a clear topcoat of finishing resin is also added. But because the patch is inside a hot tub and will not be seen, this clear coat is unnecessary.

4. **Fill** After allowing adequate drying time for the entire project, refill the hot tub. It will likely leak at the seams that you have not

fiberglassed, but the normal swelling should close those leaks. If you have fiberglassed the entire interior, the hot tub is now virtually a plastic spa.

Fiberglass spas can be patched and repaired using the same techniques. As with other repair methods, the best way to learn the nuances of preparation and application is to test the products on scrap wood or other materials before applying them at a job site. Working with fiberglass and resins is very easy and a great solution to otherwise impossible leak repairs.

Cleaning and Maintenance

RATING: EASY

Routine service for a wooden hot tub is similar to that for any other body of water, especially concerning chemical balance, vacuuming, skimming, and brushing. There are, however, some important differences about maintaining wood and prolonging its life with proper service methods (Fig. 9-18).

WOOD DECOMPOSITION

Overchlorination will strip lignin out of the wood. Lignin is the white, pulpy cellulose material that binds the organic material of the wood together. The tub will appear to be growing a white fur that will brush off and clog the filter and strainer basket with what appears to be wet newspaper.

The obvious prevention of this decomposition is to lower the doses of chlorine. Because of this problem, I don't use floaters or chlorine tablets in wooden tubs. Keep an eye on the filter and pump strainer because they will clog frequently once this decomposition has started.

The solution to this problem is to drain the tub, allow the wood to dry, and sand it down to hard, good wood. Use a coarse-grit paper (#40) to take off the worst of the material, then go back over the surface with a finer grit (#120) to smooth and seal the grain. Refill and reswell the tub.

CHEMICAL BALANCE

As the previous discussion indicates, chemical extremes can cause severe problems with wooden tubs. Since the wood itself is slightly

FIGURE 9-18   **Well-maintained redwood hot tub.** *Gordon & Grant Hot Tubs.*

acidic and the addition of bathers adds more acid, the problem with hot tubs is usually low pH, requiring a regular addition of soda ash. The acidic water will not harm the wood, but remember that it will strip metals from the components of the circulation equipment. Metal stains are not visible on the wood, so there is no early warning sign of such problems. In other words, there is no substitute for regular chemical testing and balancing of hot tub water.

Since heat and high bather activity can deplete chlorine residual between service calls, I usually premeasure a small amount of liquid chlorine and ask the customer to add it in the middle of the week. I don't use granular products in hot tubs because they can sink to the bottom before they dissolve, and at full strength will begin wood decomposition. Under no circumstances should you throw a chlorine tablet into the bottom of the tub.

ALGAE AND WOOD

If you discover an algae bloom in a hot tub, superchlorinate the water and brush all the affected surfaces. This process forces sanitizer into the grain of the wood where the algae are growing. Drain the tub and rinse with fresh water to remove excess chlorine, brushing and rinsing thoroughly. Refill the tub and treat the water normally. If you try to treat an algae bloom in a wooden tub without draining it, either you will not kill all the algae or the excess chlorine will begin wood decomposition.

Commercial Spas

Large commercial spas are not that different from residential versions, although obviously they are several times larger. Maintenance, operations, repairs, and cleaning are very similar.

One growing concern that is highlighted with commercial spas is the potential for spreading disease. The following simple precautions are also worth noting in your residential installation.

Health Issues

The focus of attention at any commercial pool or spa is health and safety. Safety is obvious, especially regarding prevention of slips and falls, chemical handling, and life saving in drowning incidents. Health concerns may be more subtle, but these days are perhaps an even greater disaster waiting to happen if proper precautions are not taken.

LEGIONNAIRES' DISEASE

Numerous outbreaks of Legionnaires' disease (a sometimes fatal respiratory ailment) have been linked to operating whirlpool spas. Outbreaks have occurred at trade shows and on cruise ships, so no installation is immune. The Centers for Disease Control (CDC) have made recommendations to prevent these tragedies, and details can be obtained by visiting the website at www.cdc.gov.

Essentially the recommendations are to clean and replace the filter regularly and to maintain a proper chlorine residual, even if the spa is only for demonstration or decorative purposes. The CDC findings in the cruise ship incident stated, "Visual examination of the filter material showed extremely heavy organic loading. This loading remained in the filter despite reports that a routine (daily) filter backwash cycle was implemented." That facility also replaced the filter cartridge annually, but inspection and common sense should have been followed instead of a predetermined schedule.

The CDC findings also noted that both UV light and ozone treatments were effective in killing the bacteria associated with Legionnaires' disease, but only in association with a residual of chlorine or bromine. The CDC also noted that copper and silver ion treatment and iodine were effective in reducing the bacteria.

Finally, the CDC recommended preventive measures in commercial spas such as hourly residual readings, daily superchlorination treatments, and weekly filter cartridge inspections. It also recommended reducing jet or bubble action in spas which tend to create aerosols of the water that can then be inhaled.

FECAL ACCIDENTS

No matter how careful you are as a commercial pool or spa manager, sooner or later a child or incontinent adult will cause a fecal accident in your facility. Many jurisdictions have written regulations for response, but here are some basics:

1. Close the pool or spa immediately and require all bathers to leave.

2. Manually remove as much fecal material as possible. If vacuuming is possible or necessary, vacuum to the sanitary sewer, not the filtration system. Disinfect the vacuum, hose, and pump before using them again.

3. Disinfect the pool or spa to a CT value of at least 9600. The CT value is the concentration of chlorine (expressed in parts per million) multiplied by the time in minutes. For example, a 20-ppm chlorine residual maintained for 8 hours (480 minutes) results in a CT value of 9600 (20 ppm × 480 minutes). Any combination of chlorine concentration and time that equals at least 9600 will be satisfactory.

4. Adjust the pH as needed to normal levels.

5. Run the filtration system for at least four complete turnovers of the water. After this cycle, backwash to the sanitary sewer. Break down and clean the filter, disinfecting it with a dilute chlorine wash (20 parts of water to 1 part of standard pool/spa chlorine).

6. If possible, it is preferable to drain the spa to the sanitary sewer and refill. Before refilling, brush the interior with a dilute chlorine wash (20 parts of water to 1 part of standard pool/spa chlorine).

7. Restart the spa and balance the water. When chlorine residuals return to 5 ppm, the spa may be reopened.

Many local codes require reporting of fecal accidents to health authorities. Of course prevention is the best measure, so keep an eye out for kids (or adults) who may be hiding diapers under their bathing suits, and keep them out of the spa in the first place.

ABS (acrylonitrile butadiene styrene)   Rigid plastic pipe similar to PVC, usually manufactured black, used for drainage systems.

A/C (alternating current)   Electric charge that flows from negative to positive, then reverses direction. The basic form of electricity used in most homes and businesses and therefore in most pool and spa equipment.

acetone   A highly flammable solvent used to clean plastic surfaces and tools.

acid   A liquid or dry chemical that lowers pH when added to water, such as muriatic acid.

acid demand   The amount of acid required (demanded) by a body of water to raise the pH to neutral (7).

acidity   A substance with a pH lower than neutral (7).

acid wash   The procedure of cleaning plaster with a solution of muriatic acid and water.

adapter bracket   The part of a pump that supports the motor and connects the motor to the pump.

aerator   A pipe vented to the atmosphere, sometimes with an adjustable volume control, added to a waterline to mix air and water prior to discharge.

air blower   *See* blower.

air relief valve   A valve on a filter that permits air to be discharged from the freeboard.

air switch   A pneumatic mechanical control device used to operate spa equipment safely. A button, located in or near the water, is depressed, sending air pressure along a plastic hose to an on/off switch.

algae   Airborne, microscopic plant life of many forms which grows in water and on underwater surfaces.

algaecides   A group of chemical substances that kill algae or inhibit their growth in water.

algaestat (algistat)   Chemical that inhibits growth of algae.

alkalinity   The characteristic of water which registers a pH above neutral (7).

ambient temperature   The average, prevailing surrounding temperature.

ammonia Natural substance composed of nitrogen and hydrogen that readily combines with free chlorine in water, forming chloramines (weak sanitizers).

amperage (amps) The term used to describe the actual strength of the electric current. It represents the volume of current passing though a conductor in a given time. Amps = watts divided by volts.

antisiphon valve A control device added to the domestic water supply line to prevent contaminated water from flowing backward into the pipe.

antisurge valve A check valve used in air blower plumbing to prevent water from entering the blower mechanism.

antivortex The property of a plumbing fitting that prevents a whirlpool effect when water is sucked through it. Used on main drain covers.

automatic gas valve The valve that controls the release of natural or propane gas to a heater. Also called the combination gas valve.

available chlorine *See* chlorine, free available.

AWG (American Wire Gauge) The standard used to specify wiring of certain thickness and capacity.

backfill Dirt, sand, or other material used to fill the gaps between a spa wall and the surrounding excavation.

backwash Process of running water through a filter opposite to the normal direction of flow to flush out contaminants.

bacteria Any of a class of microscopic plants living in soil, water, organic matter, or living beings and affecting humans as chemical reactions or viruses.

balance The term used in water chemistry to indicate that when measuring all components together (pH, total alkalinity, hardness, and temperature) the water is neither scaling or etching.

barb fitting A plumbing fitting with exterior ribs, connected by insertion into a pipe (usually flexible pipe).

base An alkaline substance.

bather Any person using the spa.

bather load The number of bathers in a spa at one time.

bayonet Refers to a lightbulb using two nipples to engage corresponding slots in the socket.

bicarbonate of soda (bicarb) A chemical used to raise pH and total alkalinity in water, also called baking soda.

bleach Colloquial term for liquid chlorine.

bleed To remove the air from a pipe or device, allowing water to fill the space.

blister An air pocket in a plaster surface.

blow bag Also called a drain flush or balloon bag, a device attached to a garden hose which expands under water or air pressure to seal an opening, forcing the water or air into that opening.

blower An electromechanical device that generates air pressure to provide spa jets and rings with bubbles.

bond beam The top of a pool or spa wall, built stronger than the wall itself to support coping stones and decking.

bonding system The wiring between electrical appliances and then to the ground to prevent electric shock in case of a faulty circuit. All appliances are grounded to the same wire.

booster pump A pump added to a spa system to create additional pressure to jets.

break down *See* tear down.

break-point chlorination The application of enough chlorine to water to combine with all ammonia (creating chloramines); then to destroy all chloramines; then leave a residual of free chlorine. The break point is, therefore, the point at which chlorine added to the water is no longer "demanded" to sanitize and is therefore available to become residual chlorine. Some references define the break point as 10 parts of chlorine for every 1 part of ammonia present in the water when the pH is between 6.8 and 7.6.

bridging The condition existing when DE and dirt close the intended gaps between filter grids in a DE filter, reducing the flow rate through the filter and reducing the square footage of filter area.

bromine (Br_2) A water-sanitizing agent. A member of the halogen family of compounds.

Btu (British thermal unit) The measurement of heat generated by a fuel. The amount of heat required to raise one pound of water one degree Fahrenheit (when at or near 39.2°F).

buffer A substance that tends to resist change in the pH of a solution.

burner tray The component in the bottom of a heater that controls the burning of gas.

bushing An internal plumbing fitting that fits (threaded or slip) into a connector fitting to reduce the internal diameter so as to accommodate a smaller pipe size than the connector fitting was designed to accept. *See* reducer.

calcium A mineral element typically found in water.

calcium hypochlorite ($Ca(OCl)_2$) A granular form of chlorine (widely produced under the brand name HTH), generally produced in a compound of 70% chlorine and 30% inert materials.

cam lock   The device that holds or releases two halves of a telepole.

cap   Plumbing fitting attached to the end of a pipe to close it completely.

cartridge   Element in a filter covered with pleats of fabric to strain impurities from water which passes through it. Generally strains out particles larger than 20 microns.

caustic   Acidic or etching in nature.

cavitation   Failure of a pump to move water when a vacuum is created because the discharge capacity of the pump exceeds the suction ability.

centrifugal force   The outward force created by an object in circular motion. The force that is used by water pumps to move water.

C frame   A type of motor housing resembling a C. It adapts to a particular style of pump.

check valve   A valve that permits flow of water or air in only one direction through a pipe.

chelating agent   Chemical compound that prevents minerals in solution in a body of water from precipitating out of solution and depositing on surfaces of the container.

chine   The portion of a stave (in a wooden hot tub) that extends below the croze.

chine joist   The sturdy plank that supports the floor of a hot tub.

chloramine   Compound of chlorine when combined with inorganic ammonia or nitrogen. Chloramines are stable and slow to release their chlorine for oxidizing (sanitizing) purposes.

chlorinator   A device that delivers chlorine to a body of water.

chlorine (Cl_2)   A substance made from salt that is used to sanitize water, killing bacteria. A member of the halogen family of compounds, chlorine is produced in gas, liquid, and granular forms.

chlorine demand   The amount of chlorine required (demanded) by a body of water to raise the chlorine residual to a predetermined level.

chlorine, free available   That portion of chlorine in a body of water that is immediately capable (available) of oxidizing contaminants.

chlorine lock   Term applied to chloramine formation, when ammonia is present in the water in sufficient quantity to combine with all available chlorine.

chlorine residual   The amount of chlorine remaining in a body of water after all organic material (including bacteria) has been oxidized, expressed in parts per million. The total chlorine residual is the sum of all free available chlorine plus any combined chlorine (chloramine).

circuit   The path through which electricity flows.

circulation system The combination of pipes, pump, and any other components through which water flows in a closed loop (from the body of water, through the components, and back to the same body of water).

clamping ring The metal band that applies pressure to and holds the lid on a filter tank or holds together two halves of a pump.

close nipple *See* nipple.

colloidal silver A compound of the metal silver, used as an algaecide.

colorimetric The name given to a chemical test procedure where reagents are added to water and change color to reflect the presence and strength of a known substance. The color is then compared to a color chart to evaluate the volume of that substance. Colorimetric tests are used, for example, to detect the presence of chlorine in water and the strength expressed in parts per million.

comparator The color chart or other device used to compare the color of a treated water sample with known values. Used in chemistry test kits.

compression fitting A plumbing connection that joins two lengths of pipe by sliding over each pipe and applying pressure to a gasket which seals the connection.

control circuit A series of safety and switching devices in a heater, all of which must be closed before electric current flows to the pilot light and automatic gas valve to ignite the heater.

coping The cap on the edge of the circumference of a pool or spa mounted on the bond beam.

corona discharge Method of producing ozone by passing electricity through oxygen and water.

coupling Plumbing fitting used to connect two lengths of pipe.

CPVC (chlorinated polyvinyl chloride) The designation of plastic pipe which can be used with extremely hot water.

croze The milled groove in the stave of a wooden hot tub into which the floorboards are inserted.

cyanurates Chlorine sanitizers combined with buffers, such as dichlor and trichlor.

cycle A complete turn of alternating current (A/C) from negative to positive and back again (*see* hertz). Also refers to a filtration period (*see* filter run).

dedicated circuit An electric circuit used only for one specific appliance.

delamination Separation or failure of the bond in layered materials such as plaster to gunite or fiberglass to resin.

diatomaceous earth (DE)   A white, powdery substance composed of tiny prehistoric skeletal remains of algae (diatoms), used as a water filtration medium in DE filters. DE filters can remove particles larger than 5 to 8 microns.

dichlor   *See* sodium dichloro-s-triazinetrione.

dielectric   An insulating material that prevents electrolysis or corrosion when applied between two surfaces of dissimilar metals.

diffuser   A housing inside a pump covering an impeller which reduces speed of the water but increases pressure in the system.

discharge   The flow of water out of a pipe or port.

dog   *See* tripper.

DPD (diethyl phenylene diamene)   The chemical reagent to detect the presence of free available chlorine in a body of water.

drain flush   *See* blow bag.

dry-fit   To assemble PVC plumbing on the job without glue to check that you have prepared the components correctly.

Dynell   Type of material used to cover filter grids.

effluent   The water discharging from a pipe or equipment.

elbow   The term for a plumbing connector fitting with a 90-degree bend.

electrolysis   The production of chemical changes (usually corrosion) to metal by passing an electric current through an electrolyte (usually water with mineral content).

electrometric   Describes a chemical test involving an electronic analysis meter.

element   A filter grid. Also the electric heat-generating rod of an electric heater.

end bell   The metal housing or cap at the end of an electric motor.

energy-efficient   Describes a specific design of pool and spa motors which have heavier wire in the windings to lower the electricity wasted from heat loss.

erosion system   A type of chemical feeder in which granular or tablet sanitizer is slowly dissolved by constant flow of water through the device.

etching   Corrosion of a surface by water which is acidic or low in total alkalinity and/or hardness.

ethylenediamine tetra-acetic acid (EDTA)   A reagent used in testing for calcium hardness, added a drop at a time until the solution turns blue. The number of drops is compared to values on a chart to evaluate the hardness of the sample.

expansion joint   The gap between the coping and the deck which allows for normal expansion and contraction of the materials, preventing damage that would otherwise result from pressure of the two against each other.

extrusion A process for making plastic or metal spa components by squeezing molten material through a precut die.

female Describes plumbing fittings or pipe with internal threads or connectors.

fiber optics Technology underlying underwater lighting devices which illuminate by sending light along thin plastic cable from a remote source.

filter Device for straining impurities from water which flows through it.

filter cycle *See* filter run.

filter filling The technique used to prime a pump by filling the filter with water and allowing it to flow backward into the pump.

filter run The time between cleanings, expressed as the total running time of the system, also called filter cycle. Care must be taken in using the term *run* or *cycle,* since some technicians mean the number of hours the system operates each day, rather than the total time between cleanings. Regionally, one term may be used for one definition and the other for the second.

FIP A female-threaded plumbing fitting.

fireman's switch An on/off control device, mounted in a time clock, that shuts off a heater 20 minutes before the time clock shuts off the circulating pump/motor. This allows the heat inside the heater to dissipate before shutdown.

flange gasket The rubber sealing ring that prevents leaks between a heater and the circulation pipes.

flapper gate The part in a check valve that swings open when water is flowing in the intended direction but swings shut when water attempts to flow backward.

flare fitting Threaded plumbing fitting that requires a widening of the pipe at one end.

flash test The method of dropping chemistry test reagents directly into pool or spa water rather than in a vial containing a test amount of that water.

flex connector A coated metal pipe with threaded fittings on each end designed to bend freely for connection in tight quarters or at odd angles. Usually used to connect gas pipe to an appliance (such as a spa heater) indoors, allowing the appliance to be moved without breaking the connection.

floater A chemical feeder system whereby a sanitizer tablet is placed in the device and it is allowed to float around the body of water. The tablet dissolves, and the sanitizer is released into the water.

flocculation   The process of adding a chemical to a body of water which combines with the suspended particulate matter in the water, creating larger particles that are more easily seen and removed from the water. Also called a clarifying agent or coagulant.

flowmeter   A device for measuring the rate of water passing through a given pipe, expressed in gallons per minute (gpm).

flow rate   The volume of a liquid passing a given point in a given time, expressed in gallons per minute (gpm).

fluorine   A little used (very costly) halogen sanitizer.

four-pass unit   A type of heat exchanger found in pool and spa heaters that directs the flow of water to be heated through the unit four times before returning it to the body of water.

freeboard   The vacant vertical area between the top of the filtration medium and the underside of the top of the filter itself.

free chlorine   Also called available chlorine. Chlorine in its elemental form, not combined with other elements, available for sanitization of water.

free joint   A colloquial expression in plumbing. When you connect several joints in a plumbing job, the free joints are the ones you can complete without committing to the end result, the ones that are still reversible.

fusible link (fuse link)   A safety device located near the burner tray of a heater, part of the control circuit. If the fuse link detects heat in excess of a preset limit, it melts and breaks the circuit to shut off the heater.

gasket   Any material (usually paper or rubber, but sometimes caulk or other pastes) inserted between two connected objects to prevent leakage of water.

gate valve   A valve that restricts water flow by raising and lowering a disk across the diameter of the pipe by means of a worm drive. *See* slide valve.

gauge   Refers to the size of an electrical wire. Heavier loads can be carried on heavier-gauge wires; however, the numbering system of wire gauges works in reverse. A 10 gauge wire, for example, is thicker than a 14 gauge wire. Also refers to a measuring device, as in pressure gauge.

gelcoat   A thin, surface, finishing coat of resin sprayed over a fiberglass job.

glazed   Refers to surfaces that have been covered with a clear, protective coating, applied under heat, such as glazed tile.

gpm   Gallons per minute. A unit of measurement.

grid   Frame covered with fabric used as a filter medium, also called septa or element.

ground fault circuit interrupter (GFI)   A circuit breaker. A sensing device that determines when electricity in a circuit is flowing through an unintended path, usually to earth, creating a hazard of electrocution. The GFI detects current variations as low as $5/1000$ of 1 amp and breaks the circuit within $1/40$ second.

gunite   A dry mixture of cement and sand mixed with water at the job site and sprayed onto contoured and supported surfaces to build a pool or spa, creating the shell.

halogens   Family of oxidizing agents including chlorine, bromine, iodine, and fluorine.

hardness   Also called calcium hardness; the amount of dissolved minerals (mostly calcium and magnesium) in a body of water.

Hartford loop   Used in air blower installations, a method of plumbing the air delivery pipe above the water level of the spa, then back below the water level. The loop prevents water from siphoning into the blower. Also the term given to the four-pass heat exchanger in some heaters.

heater   Device that raises the temperature of water using natural gas, electricity, propane, solar, or mechanical energy for fuel.

heat exchanger   The copper tubing in a heater through which water flows. The water absorbs rising heat which is generated from the burner tray below.

heat riser   A metal pipe plumbed directly to the heater to facilitate dissipation of heat before plumbing with PVC. Also called a heat sink.

hertz   Unit of measurement in A/C indicating the number of cycles per second in the current as generated. In the United States, A/C is generated at 60 hertz.

high-limit switch   A safety device used in the control circuit of heaters. When the high-limit switch detects temperatures in excess of its preset maximum, it breaks the control circuit to shut down the heater.

high-rate sand filter   A filter using sand for the filtration medium designed for flows in excess of 5 gpm but less than 20 gpm (less than 15 in some codes) per square foot. Strains impurities larger than 50 to 80 microns.

hopper   The low "bowl" portion of the spa around the main drain.

horsepower (hp)   The standard unit of measurement to denote the relative strength of a mechanical device. One horsepower is equals 746 watts, or the power required to move 550 pounds 1 foot in 1 second.

hose bib (also bib)   The faucet to which a garden hose is attached.

hydration   The process of adding moisture to a dry substance.

hydraulics   The science of water movement.

hydrochloric acid (HCl) Muriatic acid.

impeller Rotating part of a pump that creates centrifugal force to create suction. The impeller is said to be *closed* if it is shrouded (covered) on both sides of the vanes; *semi-open* if shrouded on one side, while the interior surface of the volute creates a partial shroud on the other side.

inlet *See* return.

intermittent ignition device (IID) The electronic control and switching device used in electronic ignition heaters to operate the control circuit and automatic gas valve. Often called the brain box.

J-box Short for junction box; the metal container in which wires are connected along a circuit or conduit.

jet (hydrojet) The discharge fitting through which water is returned to a spa at a high rate of speed and/or pressure.

jetted tub A bathtub fitted with hydrotherapy spa-type jets.

joint stick A paste used on threaded plumbing to prevent leaks, provided in the form of a crayonlike stick for easy application.

joist The support beam under the floor of a wooden hot tub or similar structure.

keyed shaft The shaft of a motor which has a groove for securing setscrews of the shaft extender. Used with specific designs of pumps.

kilowatt 1000 watts of electric power. Electricity is sold by the kilowatt-hour, meaning a certain fee charged for every 1000 watts delivered per hour.

laterals The horizontal filter grids at the bottom of a sand filter, installed in the underdrain.

lazy flame A natural or propane gas flame in the burner tray of a pool or spa heater which burns in slow, wavering "licks" or "stuttering," rather than the normal strong, clear blue flame burning straight upward.

leaf rake A large open net secured to a frame which attaches to a telepole. Used to skim debris from the surface of water.

lignin The white, pulpy cellulose material that actually binds the organic material of wood together. When a wooden hot tub is overchlorinated, stripping out the lignin, the tub will appear to be growing a white fur which will brush off and clog the filter.

line A wire conducting electricity. Also a run of pipe carrying water.

load An appliance that uses electricity.

main drain The suction fitting located in the lowest portion of a body of water. The principal intake for the circulation system.

main valve The flow control device in the combination gas valve of a pool or spa heater which regulates the flow of gas to the burner tray.

male Any plumbing fitting or pipe with external threads or connectors.

manifold An assembly or component that combines several other components together. A pipe fitting with several lateral outlets for connecting one pipe with others.

medium Any material used to strain impurities from water which passes through it. DE and the fabric covering a cartridge are both examples of filter media.

micron A unit of measurement equal to 1 millionth meter; also 0.0000394 inch. For example, a grain of table salt is approximately 100 microns in diameter.

MIP A male-threaded plumbing fitting.

mottling A difference in shades within a given color. The term is used mostly in plaster to denote "cloudy" patches, blotches, or streaks of uneven color.

muriatic acid Also called hydrochloric acid, this chemical is the most commonly used substance for reducing pH and total alkalinity in water.

neutral The pH reading at which the substance being measured is neither acidic nor alkaline. Neutral pH is 7.0.

niche The housing built into the wall of a pool or spa to accommodate a light fixture.

nipple Short length (less than 12 inches) of pipe threaded at each end. If the nipple is so short that the entire length is threaded, it is called a close nipple.

no-hub connector A rubber (or neoprene) plumbing fitting used to connect two lengths of pipe, attached with hose clamps or other pressure devices.

nut driver A tool like a screwdriver but fitted with a female socket that surrounds a given size of nut for loosening or tightening.

O-ring Thin rubber gasket used to create a waterproof seal in certain plumbing joints or between two parts of a device, such as between the lid and the strainer pot on a pump.

organic Any material that is naturally occurring (not manufactured), such as leaves, sweat, oil, and urine.

ORP (oxidation reduction potential) A unit of measure of sanitizer in water, measured with an electronic ORP meter.

OTO (orthotolidine) The test reagent used in detecting the presence of chlorine and, by the resulting color of the OTO, the amount of chlorine, expressed in parts per million.

ozonator A device using electricity and oxygen to create ozone and deliver it to a body of water for sanitizing purposes.

ozone (O₃)   Three molecules of oxygen, creating a colorless, odorless gas used for water sanitation.

pH   The relative acidity or alkalinity of soil or water, expressed on a scale of 0 to 14 where 7 is neutral, 0 is extremely acidic, and 14 is extremely alkaline.

phenol red (phenolsulfonephthalein)   The most widely used chemical reagent to measure the pH in a sample of water.

pi   Mathematical constant equal to 3.14.

pilot   The small gas flame that ignites the burner tray of a heater.

pilot generator   The device that converts heat from the pilot light into electricity to power a control circuit on a heater. Also called a power pile or thermocouple.

pipe dope   A paste used to prevent leaks in threaded plumbing.

pipe run   A length of pipe between two valves, connectors, or pieces of spa equipment.

plaster   A hand-applied combination of white cement, aggregates, and additives that covers the shell of a gunite pool or spa to waterproof and add beauty. Plaster can also be colored.

plug   Plumbing fitting used to close a pipe completely by inserting it (slip or threaded) into a female fitting or pipe end.

port   Opening, as in discharge port being the opening through which water flows out of a pipe or system.

portable spa   A spa that can readily be disassembled, moved, and reassembled.

power pile   *See* pilot generator.

ppm (parts per million)   The measurement of a substance within another substance; for example, 2 ounces of chlorine in 1 million ounces of water would equal 2 parts per million.

precipitate   An insoluble compound formed by chemical action between two or more normally soluble compounds. When water can no longer dissolve and hold in solution a compound, it is said to *precipitate* out of solution.

precoat   The process of applying DE to grids in a DE filter after cleaning but before restarting normal circulation and filtration.

pressure gauge   A device that registers the pressure in a water or air system, expressed in pounds per square inch.

pressure switch   A safety device in a heater control circuit that senses when there is inadequate water pressure (usually less than 2 pounds per square inch) flowing through a heater (which might therefore damage the heater) and breaks the control circuit, thereby shutting down the heater.

prime   The process of initiating water flow in a pump to commence circulation by displacing air in the suction side of the circulation system.

psi   Pounds per square inch.

pumice   A natural, soft (yet abrasive) stone substance (similar to lava rock) used to clean tiles.

pump   A mechanical device driven by an electric motor which moves water.

PVC (polyvinyl chloride)   The type of plastic pipe and fittings used most commonly in pool and spa plumbing.

radius   One half of the diameter of a circle.

reagent   A liquid or dry chemical that has been formulated for water testing. A substance (agent) that reacts (reagent) to another known substance, producing a predictable color in the water.

reducer   An external plumbing fitting that connects two pipes of different diameters. *See* bushing.

residual   The amount of a substance remaining in a body of water after the demand for that substance has been satisfied.

resin   A liquid plastic substance (with a consistency of honey) applied to fiberglass fabric to strengthen and stiffen it.

retainer   The plastic disk that fits over the top of a set of filter grids to hold them in place with the aid of a retainer rod.

retaining rod   The metal rod in the center of certain filters, on which is attached a retainer ring to hold grids in place.

return   The line and/or fitting through which water is discharged into a body of water, also called inlet.

riser   *See* heat riser. Also any vertical run of pipe.

run   Any horizontal length of pipe. *See* filter run.

sand filter   A filtration device using sand as the filter (straining) medium. *See* rapid sand filter and high-rate sand filter.

sanitizer   Any chemical compound that oxidizes organic material and bacteria to provide a clean water environment.

saturation   The point at which a body of water can no longer dissolve a mineral and hold it in solution.

scale   Calcium carbonate deposits that form on surfaces in contact with extremely hard water. Water in this condition is said to be scaling or precipitating.

seal   A device in a pump that prevents water from leaking around the motor shaft.

seal plate   The component in a pump in which the seal is situated.

sediment trap   A short length of vertical pipe attached below a run of gas supply pipe to catch any contaminants prior to the gas entering a heater.

separation tank Container used in conjunction with a DE filter to trap DE and dirt when backwashing.

sequestering agent *See* chelating agent.

shaft extender A bronze fitting added to the shaft of a motor to lengthen the shaft to accommodate the design of the pump being used. The extender is held in place with three setscrews fastened tightly against the motor shaft.

shock treatment *See* superchlorination.

skidpack A metal frame onto which is mounted the equipment needed to operate a portable spa, usually pump/motor, filter, heater, blower, and control devices. Also called a spa pack.

skimmer A part of the circulation system that removes debris from the surface of the water by drawing surface water through it.

slide valve A guillotinelike plumbing device that restricts or shuts off the flow of water in a line. In essence, a gate valve that slides a disk up or down across the flow in the pipe. *See* gate valve.

slip fitting A plumbing fitting that joins to a pipe without threads, but which slides into a prefitted space.

soda ash (sodium carbonate) (Na_2CO_3) A white, powdery substance used to raise the pH of water.

sodium bicarbonate ($NaHCO_3$) *See* bicarbonate of soda.

sodium bisulfate ($NaHSO_4$) A chemical compound used to lower the pH and total alkalinity of water (dry acid).

sodium dichloro-s-triazinetrione (dichlor) ($C_3N_3O_3Cl_2Na$) Granular, stabilized form of chlorine sanitizer, generally about 60% available.

sodium hypochlorite (NaOCl) Liquid solution containing approximately 15% chlorine.

sodium thiosulfate A chemical used to neutralize chlorine in a test sample prior to testing for pH, without which a false reading may result.

soft water Water very low in calcium and magnesium (less than 100 ppm) and therefore considered "aggressive" or "hungry" and likely to dissolve those minerals when it comes in contact with plaster.

solar panel A glass or plastic enclosure, usually approximately 4 feet by 8 feet by a few inches thick, through which water flows, absorbing heat from the sun. The basic component of a solar heating system.

spiking In water chemistry, when readings of a parameter alternately register extremely high, then extremely low.

squared A number multiplied by itself. For example, $4 \times 4 \times 4$ squared.

square flange A casing construction style of certain motors, designed to adapt to certain pumps.

stack   Vent pipe of a heater. A heater installed indoors requires such a stack, while one installed outdoors uses a flat-top vent and is called stackless.

stanchion   The vertical support pipe of a railing.

standing pilot   A heater ignition device in which the gas flame (to ignite the main burner) is always burning.

strainer basket   A plastic mesh container that strains debris from water flowing through it inside the strainer pot.

strainer pot   The housing on the intake side of a pump which contains a strainer basket and serves as a water reservoir to assist in priming.

street fitting   Any plumbing fitting that has one male end and one female end.

submersible pump   A pump/motor that can be submerged to pump out or recirculate a body of water. Also called a sump pump.

superchlorination   Periodic application of extremely high levels of chlorine (in excess of 3 ppm) to completely oxidize any organic material in a body of water (including bacteria) and leave a substantial chlorine residual.

T fitting   A plumbing fitting shaped like the letter T which connects pipes from three different sources.

tear down   To disassemble a piece of equipment for service or repair. Specifically used in reference to filter cleaning (as opposed to mere backwash or other temporary cleaning), a filter teardown means complete disassembly and cleaning. Also called breakdown.

Teflon tape   A thin fabric provided on a roll used to coat threaded fittings to prevent leaks in plumbing.

telepole   A metal or fiberglass rod that extends to twice its original length, the two sections locking together. The telepole is used with most pool and spa cleaning tools.

therm   The unit of measurement you will read on a gas bill, which is 100,000 Btu/hour of heat.

thermal overload protector   A temperature-sensitive switch on a motor that cuts the electric current in the motor when a preset temperature is exceeded.

thermistor   The sensor that sends temperature information to the thermostat.

thermocouple   *See* pilot generator. Also called thermopile.

thermostat   A part of the heater control circuit. An adjustable device that senses temperature and can be set to break the circuit when a certain temperature is reached, then close the circuit when the temperature falls below that level.

Thoroseal   A brand-name product of waterproof concrete commonly used in pool and spa repairs.

three-port valve   A plumbing fitting used to divert flow from one direction to two other directions.

time clock   An electromechanical device to automatically turn on or off an appliance at preset intervals.

titration   A chemical test method to determine the amount of a substance in a sample of water. A sample is colored, then drops of the titrant are added to the sample. When the sample turns clear or changes to another predicted color, the result of the test is determined by observing the number of drops of titrant required to create that change. An acid demand test is performed with titration, for example. The number of drops of titrant required in that test will determine the amount of acid to add to the water.

torque   Application of an amount of force against an object. Usually refers to the amount of force required to tighten a fastener to a specified requirement.

total alkalinity   The measurement of all alkaline substances (carbonates, bicarbonates, and hydroxides) in a body of water.

total dissolved solids (TDS)   The sum of all solid substances dissolved in a body of water, including minerals, chemicals, and organics.

trichloro-triazinetrione (trichlor) ($C_3N_3O_3Cl_3$)   Dry form of chlorine, produced as granule or tablet at around 90% available chlorine.

tripper   The small metal clamp that fits in the clock face of a time clock to activate the clock (on/off) at preset times. Also called a dog.

trisodium phosphate   Detergent used to clean filter grids and cartridges, breaking down oils which acid washing alone cannot. Commercially sold as TSP.

turbidity   Cloudiness.

turnover rate   The amount of time required for a circulation system to filter 100% of the water in a particular body of water.

underdrain   The assembly in the bottom of a sand filter that connects laterals and plumbing for water filtration.

UL (Underwriters Laboratories)   The agency that sets standards for and certifies the safety of electrical and other mechanical devices.

union   Plumbing fitting connecting two pipes by means of threaded male and female counterparts on the end of each pipe.

vacuum   Device used to clean the underwater surfaces of a pool or spa by creating suction in a hose line.

valve   A device in plumbing that controls the flow of water.

volt The basic unit of electric current measurement expressing the potential or "pressure" of the current. Volts = watts ÷ amps.

volute Part of the pump. The volute is the housing that surrounds the impeller and diffuser, channeling the water to a discharge pipe.

watt The way appliances are measured for power consumption. One watt is equal to the "volume" of 1 amp delivered at the "pressure" of 1 volt. Watts = amps × volts.

weir The barrier in a skimmer over which water flows. A floating weir raises and lowers its level to match the water level in a pool or spa. Another type is shaped like a barrel and floats up and down inside a skimmer basket.

winterizing The process of preparing a pool or spa to prevent damage from freezing temperatures and other harsh weather conditions.

SPA AND HOT TUB RESOURCE GUIDE

Owning a spa or hot tub can be a trouble-free experience if you know where to go for help when things do go wrong. With modern technology, good advice is often only a click away at a convenient website. Trade magazines and related publications provide even more in-depth support for the service professional and dedicated do-it-yourselfer.

Of course, the place where you bought your spa or hot tub is another great place for advice, as is the pool and spa retailer who sells supplies. You may be tempted to buy chemicals and supplies from a grocery store or discount chain, but buying from a spa retailer will save you more money in the long run, because the professional advice that comes with the purchase will prevent costly mistakes in spa maintenance and repair.

Websites

The websites mentioned below were used in the writing of this book, but are by no means a complete list of references that might help you. Many of the websites feature both spas and spa equipment. Almost all sites offer FAQs (frequently asked questions) about spas and/or hot tubs, water maintenance, and links to other sites. My notes highlight especially good aspects of these sites, but every one of them has more to offer than could possibly be indexed here.

It's also worth mentioning that some websites change frequently, while others should be updated more often than they are. Look around and even revisit sites frequently to see how they changed. Also look at links on each website, because many sites are themselves great references to other helpful sites.

Finally, I have added an asterisk to each website that provides the best prices for various spas, supplies, and parts. This will allow you to make realistic cost estimates of spa components, equipment, and upgrades, and to estimate the cost of a job, whether you do it yourself or call a pro. Have fun!

aqua-flo.com	User-friendly pool/spa equipment specifications, photos, pdf format downloads
aquaticpartsco.com*	Buy spa parts cheap!
aquamagazine.com	Good links to industry resources
balboadirect.com	Controls, skidpacks, components; easy to search with good technical downloads
bradfordproducts.com	Unique stainless steel spa with tile accents
colemanspas.com	Good interactive guide to choosing the right spa
coloradospas.com	Variety of spas, hot tubs, and installation ideas; great "care and feeding" section
diamondspas.com/spas	Unique stainless steel jetted tubs and spas
dreammakerspas.com	Affordable portable spas; great online repair videos
eere.energy.gov/consumerinfo	Very creative, useful site about saving water and energy in pools and spas
gordonandgrant.com	One of the finest makers of redwood hot tubs; great site with lots of specialized information
gpspool.com	Good list of manufacturers; good catalogs and parts information
gnht.com	Good buyer's guide to selecting a hot tub; innovative spa/hot tub units
hurlcon.com.au	Australian pool/spa equipment manufacturer of excellent products and technical information

intermatic.com	Everything you could want to know about timers
jacuzzi.com	Great online owner's manuals and installation guides
jandy.com	Good selection on pool/spa equipment, especially Jandy/Laars heaters, good technical information
magnetek.com	Great technical library once you get through to the consumer pool/spa section; cool "Energy Savings Predictor" software
masterplumbers.com/ utilities/converter	Converts measurements from metric to U.S. standard and back; good plumbing supplies information
masterspas.com	Innovative line of spas and accessories; lots of tips and insights into ownership
medallionpools.com/spas	Unique lightweight aluminum spa with vinyl liner and redwood trim
pentairpool.com	Very useful "answer pool" section, including a neat pool/spa capacity calculator; good literature and downloads
poolspa.com	Limited links to other manufacturers
poolspanews.com*	Great searchable site with lots of technical help and links
poolproducts.com*	Good directory of tips, products, spas
poolsupplies.com*	Great site representing variety of spa types, supplies, information; sells unique Dragonfire hot tubs
raypak.com	Easy to download specification sheets on pool and spa heaters
solar-tec.com	Good user-friendly site for introduction to solar; pricing; focus on pools, but equally useful for spa installations

spaparts.com*	Very easy to navigate, good closeups of parts; unique product lines
spatop.com	Everything you could ever need in spa covers and clever accessories
starite.com	Good site with easy-to-get owner's manuals and great chart that tells you how long each will take to download at your modem speed
sundancespas.com	Downloads of owner's manuals with good technical information
suntreksolar.com	Great do-it-yourself products and step-by-step installation guide

Trade Publications for the Service Professional

Aqua (ask for their excellent annual *Buyers' Guide* when you subscribe)
4130 Lien Road
Madison, WI 53704
(608) 249-0186
aquamagazine.com

Pool & Spa News (ask for their excellent annual "Directory Issue" when you subscribe)
4160 Wilshire Blvd.
Los Angeles, CA 90010
(323) 964-4800
poolspanews.com

Service Industry News
P.O. Box 5829
San Clemente, CA 92674
(949) 366-9981

Swimming Pool & Spa Age
6151 Powers Ferry Road
Atlanta, GA 30339
(770) 955-2500
poolspamag.com

INDEX

ABOUT THE AUTHOR

Terry Tamminen has spent 30 years in the pool and spa industry in Florida, California, and west Africa. He is the author of *The Ultimate Pool Maintenance Manual* (McGraw-Hill, 2001) and *The Ultimate Guide to Above-Ground Pools* (McGraw-Hill, 2004).

His passion for clean water extends to the marine environment, where he works with the International Waterkeeper Alliance to restore and protect coastal waters in the United States and abroad.

In 2004, Mr. Tamminen was appointed to serve as the Cabinet Secretary for Governor Arnold Schwarzenegger in California. He is also a Shakespearean actor and scholar, a private pilot, and an avid scuba diver.